T0136883

Studies in Big Data

Volume 55

Series editor

Janusz Kacprzyk, Polish Academy of Sciences, Warsaw, Poland

The series "Studies in Big Data" (SBD) publishes new developments and advances in the various areas of Big Data- quickly and with a high quality. The intent is to cover the theory, research, development, and applications of Big Data, as embedded in the fields of engineering, computer science, physics, economics and life sciences. The books of the series refer to the analysis and understanding of large, complex, and/or distributed data sets generated from recent digital sources coming from sensors or other physical instruments as well as simulations, crowd sourcing, social networks or other internet transactions, such as emails or video click streams and others. The series contains monographs, lecture notes and edited volumes in Big Data spanning the areas of computational intelligence including neural networks, evolutionary computation, soft computing, fuzzy systems, as well as artificial intelligence, data mining, modern statistics and Operations research, as well as self-organizing systems. Of particular value to both the contributors and the readership are the short publication timeframe and the world-wide distribution, which enable both wide and rapid dissemination of research output.

** Indexing: The books of this series are submitted to ISI Web of Science, DBLP, Ulrichs, MathSciNet, Current Mathematical Publications, Mathematical Reviews, Zentralblatt Math: MetaPress and Springerlink.

More information about this series at http://www.springer.com/series/11970

Katarzyna Tarnowska · Zbigniew W. Ras ·
Lynn Daniel

Recommender System for Improving Customer Loyalty

Katarzyna Tarnowska
Department of Computer Science
San Jose State University
San Jose, CA, USA

Lynn Daniel
The Daniel Group
Charlotte, NC, USA

Zbigniew W. Ras
Department of Computer Science
University of North Carolina
Charlotte, NC, USA

Polish-Japanese Academy
of Information Technology
Warsaw, Poland

ISSN 2197-6503   ISSN 2197-6511   (electronic)
Studies in Big Data
ISBN 978-3-030-13440-2   ISBN 978-3-030-13438-9   (eBook)
https://doi.org/10.1007/978-3-030-13438-9

Library of Congress Control Number: 2019932683

This Springer imprint is published by the registered company Springer Nature Switzerland AG
The registered company address is: Gewerbestrasse 11, 6330 Cham, Switzerland

Preface

This book presents a novel data-driven approach to solve the problem of improving customer loyalty and customer retention. The data mining concepts of action rules and meta-actions are used to extract actionable knowledge from customer survey data and build a knowledge-based recommender system (CLIRS—Customer Loyalty Improvement Recommender System). Also, a novel approach to extract meta-actions from the text is presented. So far, the use of meta-actions required a pre-defined knowledge of the domain (i.e., medicine). In this research, an automatic extraction of meta-actions is proposed and implemented by applying Natural Language Processing and Sentiment Analysis techniques on the customer reviews. The system's recommendations were optimized by means of implemented mechanism of triggering optimal sets of action rules. The optimality of recommendations was defined as maximal Net Promoter Score impact given minimal changes in the company's service. Also, data visualization techniques are proposed and implemented to improve understanding of the multidimensional data analysis, data mining results, and interacting with the recommender system's results.

Another important contribution of this research lies in proposing a strategy for building a new set of action rules from text data based on sentiment analysis and folksonomy. This new approach proposes a strategy for building recommendations directly from action rules, without triggering them by meta-actions. The coverage and accuracy of the opinion mining were significantly improved within a series of experiments, which resulted in better recommendations. Therefore, the research presents a novel approach to build a knowledge-based recommender system whenever only text data is available.

San Jose, USA
Charlotte, USA/Warsaw, Poland
Charlotte, USA

Katarzyna Tarnowska
Zbigniew W. Ras
Lynn Daniel

About the Book

In this book, data mining and text mining techniques are a proposed approach to the problem of customer loyalty improvement. The built solution is an automated data-driven and user-friendly recommender system based on actionable knowledge and sentiment analysis.

The system proved to work in real settings and its results have already been discussed with the end business users. Its main value lies in suggesting and quantifying the effectiveness of the course of strategic actions to improve the company's growth potential. Another strength of the approach is that it works on the overall knowledge in the industry, which means that worse-performing companies can learn from the knowledge and experience of their better-performing competition.

In this book, we introduce the problem area and describe the dataset on which the work has been done. We present the background knowledge and techniques necessary to understand the built solution, as well as the current state of the art applications in the researched area.

Further, we describe all the work that already has been done within this research project, including: visual techniques applied to enhance interactiveness and friendliness of the system, experiments on improving the knowledge miner and finally the new architecture and the implementation of the system that works solely on the text customer feedback comments. Lastly, we focus our research on finding and testing new ways of improving the algorithm for natural language processing of text comments and we present the results.

Within the work done, we identified new topics that need further research and improvement, which will become the focus towards the future directions in this research.

Contents

List of Figures

List of Tables

Chapter 1
Introduction

1.1 Why Customer Experience Matters More Now?

Customer experience! These are two words mentioned frequently in business conversations as well as the business press. To illustrate just how frequently, searching customer experience in Google yields 1.3 billion results! By comparison, a search for CRM results in only 157 million results; employee engagement 164 million; and ERP (enterprise resource planning software) 168 million. Why the strong interest in customer experience? Simply, a need to find more and better ways to differentiate the products and services offered by a company. Consider some history. In the 1990s, industrial managers spent a lot of time on product innovation. Having a new product with innovative features was a proven way to win customers. The construction equipment industry provides a great example. Prior to the early 1970s, bulldozers had manual transmissions. In the early 1970s, JCB first introduced a bulldozer with a hydrostatic transmission. This type of transmission dramatically improved machine and operator productivity among other things. No longer was it necessary to stop, engage a clutch and shift gears to change direction or change the speed. Now, these changes could be done with a joy stick. Companies rapidly developed such offerings and now, nearly all bulldozers sold now have either a hydrostatic or automatic transmission. While there are differences in the transmissions offered by the various manufacturers, providing a product that did not require manual shifting became a required feature. In addition to illustrating the importance of product improvement as a key business strategy, it also shows the limits of such a strategy. In a world where most every supplier offers a similar set of product features, innovation becomes a less potent tool for differentiation. Customer experience was seldom in many strategic planning sessions discussed formally. Unless the customer experience was just awful, managers assumed it was adequate. In short, poor service could, in some cases, hinder a deal but great customer service was not seen as a tool to create more loyal customers and cause them to talk with other potential customers. To illustrate

© Springer Nature Switzerland AG 2020
K. Tarnowska et al., *Recommender System for Improving Customer Loyalty*,
Studies in Big Data 55, https://doi.org/10.1007/978-3-030-13438-9_1

the thinking of the time, consider the work of one of the leading business strategy thinkers, Michael Porter. In 1979, he wrote an article entitled "The Five Competitive Forces that Shape Strategy" (Harvard Business Review, January 2008). In the article, he outlined five forces that help to shape business strategy. To compete against these forces he suggested three broad strategies: cost leadership; differentiation; and focus. When he discussed the differentiation strategy, it was largely about product differentiation and/or cost superiority. The above is not a criticism of the profound impact that Michael Porter's strategic thinking has had on many business leaders. Rather, it illustrates the lack of attention paid to customer service as a strategic differentiator. In the mid-2000s, this calculus changed, for two big reasons. First, design and engineering capability and capacity increased throughout the world. For example, the number of science and engineering degrees awarded in China increased from 359,000 in 2000 to 1.65 million in 2014 (article from Science Policy News of the American Institute of Physics, January 2018. The article summarizes the National Science Board's biennial Science and Engineering Indicators research). India has also been producing large numbers of engineers and scientists as has Europe, Japan and the US. Consider the current race between China and the US to build the 5G wireless network of the future. Huawei Technologies, a Chinese company, had one of its components named as a critical component to the 5G system and the developer of that technology is a Turkish-born scientist named Erdal Arikan (Wall Street Journal, The 5G Race: China and U.S. Battle to Control World's Fastest Wireless Internet, September 9, 2018). 5G Race: This huge increase in intellectual capital meant that product innovation was no longer primarily found in the US, Japan or Europe. New and innovative products began appearing from a wide variety of countries. Couple a first-rate product design with competitive production costs and you have a strong value proposition. Consider what Hyundai and Kia (now part of the same company), South Korean automotive companies have done. While some may argue that the products designs are uninspiring, both offer a line of automobiles that are well-engineered and attractively priced. US sales are growing. Second, during this period, the internet came to the fore. The dotcom bubble in the late 1990s resulted in a massive overinvestment in high-speed fiber optic cable. This high-speed capacity made it easier to get information about products and companies anywhere in the world. Thomas Friedman wrote about it most eloquently in the World is Flat, when he said it leveled the playing field. While product differentiation was (and is) an important strategy for any company, it has become one of several important strategies and not the primary one. With the advent of more and stronger technical skills in many parts of the world, a differentiation strategy depending almost exclusively on product innovation is less defensible than in the past. With the plethora of information of all types, potential buyers can find out if a product's technical superiority is really all that superior! Critically, they can find out what existing users think about a product and the service it receives. Changing customer expectations have also played a critical role in making customer experience more important in the business-to-business arena. The business-to-consumer arena played a role in this as did the increasing complexity of many of the products and services sold in the business-to-business arena. How did the B-to-C market influence B-to-B? One big example is Amazon. The company made it far easier to find products, get user reviews of those products and then order them. Consumers liked it (as do shareholders since Amazon now has a market cap of over one trillion dollars). There are many other examples in addition to Amazon but suffice it to say, that they made it quicker and easier to get consumer products? Those same B-to-B buyers were also consumers of a range of products they purchased online. They realized how easy and frictionless it was, which influenced their expectations when it came to purchasing products and services for the businesses in which they worked. To illustrate, visit most any industrial products company website and they will have a link to a parts store or, in come cases, the option to order online. Rather than picking up the phone or meeting with a salesperson, customers want the ease of ordering online, where possible, much in the same way they do when they order a product from Amazon. Increasing product complexity has also created an increased need for improving the customer experience. Electronic technology of all types is increasingly integrated into products. While the benefits of this technology are quite significant there are increased product support needs. The typical higher horsepower farm tractor now comes with a climate-controlled cab that contains two or three electronic displays. These displays are monitoring tractor performance (e.g., engine, transmission, etc.), attachment functions (e.g., planter placing seeds accurately), and another that shows the GPS location so that the operator can turn the tractor at the end of the row, get it in the correct position and let the GPS system drive it to the end of the row (the operator is only monitoring). Compare this with a tractor of 20 years ago that may have had a simple cover for operator protection from the elements and no monitors. A very different picture than what is happening today. This technology is a real productivity booster but learning how to use it and keeping it maintained require significant after-sale support. When a customer spends hundreds of thousands of dollars on a piece of equipment, keeping it running and running efficiently is critical to realizing the desired return on investment.

1.2 Top (and Bottom) Line Reasons for Better Customer Experience

The previous few paragraphs outline key macro reason for why a product differentiation strategy became less strategically potent. At the micro level, managers discovered that having better customer experience pays in many ways. For example:

1. A research project showed that the higher the satisfaction the greater increase in sales. Customers that had the best past experiences spent a 140% more than customers with lower satisfaction ratings (The Value of Customer Experience Quantified, Harvard Business Review, August 1, 2014).

2. Internal research conducted by The Daniel Group shows that more satisfied customers actively refer. In the farm equipment market, they found that about 40% of farmers indicated they gave a referral for a dealer in the past 6 months. However, over 90% of the referrals came from the most satisfied customers.
3. A study of publicly traded companies by Watermark Consulting showed that from 2007 to 2013, those companies with better customer service generated a total return to shareholders that was 26 points higher than the S and P 500.

Managers today are learning that creating meaningful differentiation is about offering value in different ways and not just one or two. AGCO, a global manufacturer of agricultural equipment just recently introduced its new row-crop tractor (http://www.challenger-ag.us/products/tractors/1000-series-high-horsepower-row-crop-tractors.html). The tractor offers an innovative design with many benefits to its users (e.g., buy one tractor that can do the job of two) and it is proving to be a very high-quality product. Simultaneously, AGCO is working to improve the product support its dealers are providing buyers. Based on initial results, this multi-pronged strategy is working. The product is getting good reviews. Any issues that customers are having in the field are being promptly handled by AGCO and its dealer network. An improved customer experience matters for many financial reasons, as noted previously. If one of a company's strategies is built around product differentiation, then, as the AGCO illustration shows, a better service experience enhances the chances of success for product differentiation. What problems does the recommender system address? The recommender system addresses several important issues. The most important are:

1. Provides a decision framework so managers can understand which action or set of actions are likely to have the greatest impact on the Net Promoter Score (NPS). Managers often know they need to improve customer experience but are often lost as to where to begin. If managers need to improve inventory turns, as an example of another exercise they may do, there are pieces of information that can help guide and inform decision-making (e.g., identify inventory items with low turns and high investment). This information helps them to identify priorities for action. When faced with improving customer experience, there are few similar pieces of information to inform the decision. For example, if a difficult-to-use phone system is raised by customers, what happens to likelihood to recommend if the phone system is improved? The recommender system can provide insights as to what the likely NPS improvement is likely to be if the phone systems are improved. This improves managerial decision-making.
2. Results are based on multiple clients. The data mining techniques deployed in the recommender system allow for learning to be gained based on the experience of other users, without sharing proprietary information. This dramatically strengthens the power of the system.

3. Strengthens traditional text mining options. Text mining can be useful to identify the frequency with which topics are mentioned and the sentiment associated with the topics. The recommender system allows users to see specific anonymous comments associated with actual customers. Studying these comments can provide very granular insights into steps that can be taken to improve customer experience. In addition to quantifying the potential impact on NPS of various actions, it allows users to better understand the specific things that when done, improve NPS and, therefore, the overall customer experience.
4. Provides a sensitivity analysis. In some cases, certain actions or sets of actions are more easily implemented than others. The recommender system allows managers to weight these actions to determine which ones have more impact. For example, one action may be to conduct training to improve communication. While it may be shown to positively impact NPS, it may not be very practical in large and sprawling organizations. The system allows the various sets of actions to be weighted for feasibility, which enhances the managerial decision-making process.

1.3 What is Next?

We worked on this system for a few years and it was a great learning experience for all of us. But there is more to be done. Our work showed that it is possible to create a rigorous framework by which to analyze various actions to improve NPS. This helps improve decision-making for managers tasked with improving service experience. This framework needs to be further enhanced and strengthened to identify even better ways to derive meaning from textual comments and more deeply understand the impact on NPS of various actions. We also need to learn how this approach applies to other markets and industries. Our suspicion is that it is very applicable. The firm we collaborate with works in a variety of industrial markets and we often see similar issues no matter the industry. For example, poor communication is one of those ubiquitous problems that frequently shows up as something that negatively impacts customer experience in a variety of industries. It would be interesting and useful to see how customers in different industries respond to similar actions to improve communication.

1.4 Final Observations

Customer experience is a very robust element in the strategic arsenal of a company. While product innovation is important, it is also becoming more challenging to implement due to greater information flow and a much larger pool of creative talent throughout the world. However, a company that provides easy, reliable customer service is a tough competitor. A company with great customer experience has a very defensible strategy that is difficult to challenge. There is still much to be learned about effective and powerful ways to improve customer experience. The recommender system provides some powerful insights into how this can be done.

Chapter 2
Customer Loyalty Improvement

2.1 Introduction to the Problem Area

Nowadays most businesses, whether small-, medium-sized or enterprise-level organizations with hundreds or thousands of locations collect their customers feedback on products or services. A popular industry standard for measuring customer satisfaction is called "Net Promoter Score"[1] [1] based on the percentage of customers classified as "detractors", "passives" and "promoters" (see Fig. 2.1). Promoters are loyal enthusiasts who are buying from a company and urge their friends to do so. Passives are satisfied but unenthusiastic customers who can be easily taken by competitors, while detractors are the least loyal customers who may urge their friends to avoid that company. The total Net Promoter Score is computed as %Promoters -%Detractors. The goal here is to maximize NPS, which in practice, as it turns out, is a difficult task to achieve, especially when the company has already quite high NPS.

Most executives would like to know not only the changes of that score, but also why the score moved up or down. More helpful and insightful would be to look beyond the surface level and dive into the entire anatomy of feedback.

The main problem to solve is to understand difference in data patterns of customer sentiment on a single client personalization level, in years 2011–2016. The same it should enabled to explain changes, as well as predict sentiment changes in the future. Actionable knowledge is needed for the business for designing the strategic directions that would help drive customer loyalty improvement.

---

[1]NPS®, Net Promoter®and Net Promoter®Score are registered trademarks of Satmetrix Systems, Inc., Bain and Company and Fred Reichheld.

© Springer Nature Switzerland AG 2020
K. Tarnowska et al., *Recommender System for Improving Customer Loyalty*,
Studies in Big Data 55, https://doi.org/10.1007/978-3-030-13438-9_2

Fig. 2.1 The concept of Net Promoter Score as a way to quantify and categorize customer satisfaction

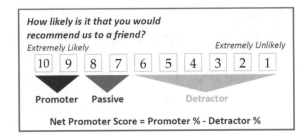

2.2 Dataset Description

The chosen dataset is related to a research project conducted in the KDD Lab at UNC-Charlotte in collaboration with a consulting company based in Charlotte. The company collects data from telephone surveys on customer's satisfaction from re-pair service done by heavy equipment repair companies (called clients). There are different types of surveys, depending on which area of customer satisfaction they focus on: Service, Parts, Rentals, etc. The consulting company provides advisory for improving their clients' Net Promoter Score rating and growth performance in general. Advised clients are scattered among all the states in the US (as well as Canada) and can have many subsidiaries. There are above 400,000 records in the dataset in total (years 2011–2016), and the data is kept being collected. The dataset (Fig. 2.2) consists of features related to:

1. Clients' details (repair company's name, division, etc.);
2. Type of service done, repair costs, location and time;
3. Customer's details (name, contact, address);
4. Survey details (timestamp, localization) and customers' answers to the questions in survey;
5. Each answer is scored with 1–10 (optionally textual comment) and based on total average score (*PromoterScore*) a customer is labeled as either Promoter, Passive or Detractor of the given client.

The data is high-dimensional with many features related to particular assessment (survey questions) areas, their scores and textual comments. The consulted clients as

Client attributes			Customer attributes			Service attributes		Survey attributes and questions (customer experience on client's service)				NPS Status
ID	Name	Adress, ...	Name	Location	...	Time	Cost ,...	Q1 (score)	Q2 (score)	Q... (score)	QN (score)	Promoter
1												Passive
2												Detractor

Fig. 2.2 Illustration of the NPS dataset structure—the features and the decision attribute

well as surveyed customers are spread geographically across United States. Records are described with temporal features, such as *DateInterviewed, InvoiceDate* and *WorkOrderCloseDate*.

2.3   Decision Problem

The first goal is to find characteristics (features) which most strongly correlate with *Promoter/Detractor* label (*PromoterScore*), so that we can identify areas, where improvement can lead to changing a customer's status from Detractor to Promoter (improvement of customer's satisfaction and client's performance). Identifying these variables (areas) helps in removing redundancy.

So far correlation analysis has considered only global statistics, however global statistics can hide potentially important differentiating local variation. The problem is multidimensional as the correlations vary across space, with scale and over time.

The intermediate goal is also to explore the geography of the issue and use interactive visualization to identify interdependencies in multivariate dataset. It should support geographically informed multidimensional analysis and discover local patterns in customers' experience and service assessment. Finally, classification on NPS should be performed on semantically similar customers (similarity can be also based on geography). A subset of most relevant features should be chosen to build a classification model.

2.4   Problem Area

The following subsections present the most important problem areas identified within the research.

2.4.1   Attribute Analysis

The first problem that needs to be solved is to find out which benchmarks are the most relevant for Promoter Status. There is also a need to analyze how the importance of benchmarks changed over years for different clients (locally) and in general (globally), and additionally how these changes affected changes in Net Promoter Score, especially if this score deteriorated (which means customer satisfaction worsened). There is a need to identify what triggered the highest NPS decreases and the highest NPS growths.

The business questions to answer here are:

- What (which aspects) triggered changes in our Net Promoter Score?

- Where did we go wrong? What could be improved?
- What are the trends in customer sentiment towards our services? Did more of them become Promoters? Passives? Detractors? Did Promoters become Passives? Promoters become Detractors? If yes, why?

The problem with the data is that the set of benchmarks asked is not consistent and varies for customers, clients and years. Customer expectations change as well. Therefore, one has to deal with a highly incomplete and multidimensional data problem.

2.4.2   Attribute Reduction

The consulting company developed over 200 such benchmarks in total (223 in 2016), but taking into considerations time constraints for conducting a survey on one customer it is impossible to ask all of them. Usually only some small subsets of them are asked in a survey. There is a need for a benchmark (survey) reduction, but it is not obvious which one of them should be asked so that to obtain the most insightful knowledge. For example, consulting aims to reduce the number of questions to the three most important, such as "Overall Satisfaction", "Referral Behavior" and "Promoter Score", but it has to be checked if this will not lead to significant knowledge loss about customer satisfaction problem. There is a need to know which benchmarks can/cannot be dropped in order to decrease a knowledge loss. For example, in years 2014–2015 some questions were asked less frequently because questionnaire structure changed and survey shortened for some clients. There is a need for analysis regarding how these changes in the dataset affect the previously built classification and recommender system model.

2.4.3   Customer Satisfaction Analysis and Recognition

The second application area is tracking the quality of the knowledge base being collected year by year, especially in terms of its ability to discern between different types of customers defined as *Promoters*, *Passives* and *Detractors*. The main questions business would like to know the answers to are:

- What (which aspect of service provided) makes their customers being Promoters or Detractors?
- Which area of the service needs improvement so that we can maximize customer satisfaction?

For every client company the minimum set of features (benchmarks) needed to classify correctly if a customer is a Promoter, Passive or Detractor should be identified. The strength of these features and how important a feature is in recognition process needs to be determined. However, answering these questions is not an easy task, as the problem is multidimensional, varies in space and time and is highly dependent on the data structure. Sufficient number of customer feedback on various aspects must be collected and analyzed in order to answer these questions. Often human abilities are not sufficient to analyze such huge volume of data in terms of so many aspects. There is a need for some kind of automation of the task or visual analytics support.

2.4.4 Providing Recommendations

The main goal of this research work is to support consulting business with recommendations (recommendable sets of actions) to their clients (repair companies), so that they can improve their NPS. The items must be evaluated in terms of some objective metrics.

Besides, the recommendation process needs to be more transparent, valid and trustworthy. Therefore, the need to visualize an algorithm process which leads to generating a recommendation output. The end user must be able to understand how recommendation model works in order to be able to explain and defend the model validity. Visual techniques should facilitate this process.

Reference

1. SATMETRIX. Improving your net promoter scores through strategic account management. http://info.satmetrix.com/white-paper-download-page-improving-your-net-promoter-scores-through-strategic-account-management. Accessed: 2017-04-26.

Chapter 3
State of the Art

In this chapter, different types of available information technology solutions support-ing customer relationship management as well as collecting the customer feedback, are discussed, with the focus on the new generation on intelligent decision support and recommender systems.

3.1 Customer Satisfaction Software Tools

Horst Schulz, former president of the Ritz-Carlton Hotel Company, was famously quoted as saying: "Unless you have 100% customer satisfaction…you must im-prove". Customer satisfaction software helps to measure customers' satisfaction as well as gain insight into ways to achieve higher satisfaction. *SurveyMonkey* [1] is the industry leading online survey tool used by millions of businesses across the world. It helps to create any type of survey, but it also lacks features with regard to measur-ing satisfaction and getting actionable feedback. *Client Heartbeat* [2] is another tool built specifically to measure customer satisfaction, track changes in satisfaction lev-els and identify customers 'at risk'. *SurveyGizmo* [3] is another professional tool for gathering customer feedback. It offers customizable customer satisfaction surveys, but it also lacks features that would help to intelligently analyze the data. *Customer Sure* [4] is a tool that focuses on customer feedback: facilitates distribution of cus-tomer surveys, gathering the results. It allows to act intelligently on the feedback by tracing customer satisfaction scores over time and observe trends. *Floqapp* [5] is a tool that offers customer satisfaction survey templates, collects the data and puts it into reports. *Temper* [6] is better at gauging satisfaction as opposed to just being a survey tool. Similar to *Client Heartbeat*, Temper measures and tracks customer satisfaction over a period of time.

These types of tools mostly facilitate design of surveys, however, offer very lim-ited analytics and insight into customer feedback. It mostly confines to simple trend analysis (tracing if score increased or decreased over time). The *Qualtrics* Insight Platform [7] is a leading platform for actionable customer, market and employee

© Springer Nature Switzerland AG 2020

K. Tarnowska et al., *Recommender System for Improving Customer Loyalty*,
Studies in Big Data 55, https://doi.org/10.1007/978-3-030-13438-9_3

insights. Besides customers' feedback collection, analysis and sharing it offers extensive insight capabilities, including tools for end-to-end customer experience management programs, customer and market research and employee engagement.

3.2 Customer Relationship Management Systems

CRM is described as "managerial efforts to manage business interactions with customers by combining business processes and technologies that seek to understand a company's customers" [8], i.e. structuring and managing the relationship with customers. CRM covers all the processes related to customer acquisition, customer cultivation, and customer retention. CRM also involves development of the offer: which products to sell to which customers and through which channel. CRM seeks to retain customers and design marketing campaigns. Sometimes CRM strategy is incorporated into other enterprise systems. An enterprise data warehouse has become a critical component of a successful CRM strategy [9]. Data mining techniques in this area are useful for extracting marketing knowledge and further supporting marketing decisions. The CRM systems must analyze the data using statistical tools and data mining. There are two critical components of marketing intelligence: customer data transformation and customer knowledge discovery.

3.3 Decision Support Systems

A DSS is an interactive computer-based system designed to help in decision making situations by utilizing data and models to solve unstructured problems [8].

  The aim of DSSs is to improve and expedite the processes by which management makes and communicates decisions—in most cases the emphasis in DSSs is on increasing individual and organizational effectiveness. DSS in general can improve strategic planning and strategic control. Research indicates data-driven or data-informed organizations improve decision-making, increase profitability and drive innovation. As strategic planning requires large amount of information, the only effective way to manage large amounts of information is with information technologies. Proper integration of DSSs and CRM presents new opportunities for enhancing the quality of support provided by each system.

3.4 Recommender Systems

Most recommender systems were applied in e-commerce settings, supporting customers in online purchases of commodity products such as books, CDs. Idea of applying recommender system in the area of strategic business planning seems to be quite novel, but the general principle of recommender systems is applicable to multiple and diverse environments.

Recommender systems, particularly using collaborative techniques, aim to predict the preferences of an individual (user/customer) and provide suggestions of further resources or items that are likely to be of interest. Formally, recommender systems are defined as programs that attempt to recommend the most suitable *items* (products or services) to particular *users* (individuals or businesses) by predicting a user's interest in an item based on related information about the items, the users and the interactions between items and users [10].

3.4.1 Recommender Systems for B2B

While most research has been focused on applying the method to help the customers in Business-To-Customer (B2C) electronic commerce, the participants in Business-to-Business (B2B) market can also get useful assistance from the recommender system. From the view of company, the recommender system should not only be able to help the buyers find their expected products and services, but also to help the sellers understand their customers better. Moreover, the system should also assist both the seller and buyer to setup strategy relationship in the long-term cooperation. For B2B relationship information of interest would be contracts; product prices, quality of service, payment terms. Past studies have shown that consumer satisfaction of B2B is lower than that of B2C, which indicates that the enterprises in B2B market did not understand their customers profoundly and make efficient response to their needs. Applying recommender system technology to B2B can bring numerous advantages: streamline the business transaction process, improve the customer relationship management, improve customer satisfaction and make dealers understand each other better. B2B participants can receive different useful suggestions from the system to help them do better business. Also, recommender system in B2B can be linked with the enterprise's back-end information system and augment the company's marketing. The system can analyze sales history and customers' comments and give advices on marketing issues, such as: how to improve the product, the customers' purchase trends, etc.

One of the proposed definition of a recommender system for B2B e-commerce was given in [11]: "a software agent that can learn the interests, needs, and other important business characteristics of the business dealers and then make recommendations accordingly. The systems use product/service knowledge—either hand-coded knowledge provided by experts knowledge learned from the behavior of consumers—guide the business dealers through the often overwhelming task of locating transactions their companies will like".

One of the important issues that must be considered in system is the output design and the presentation of the output—e.g. how to deliver the recommendations to the users, should the system explain the results to the user, etc. The most important contribution of explanations is to allow the system users to make more informed and accurate decisions about which recommendations to utilize.

3.4.2 Types of Recommender Systems

Current generation of recommendation methods can be divided into four groups:

- collaborative,
- content-based,
- knowledge-based,
- hybrid.

In comparison to other traditional information system tools and techniques, such as databases or search engines, the study of recommender systems is relatively new. It emerged as an independent research area in the mid-1990s, but interest in it increased dramatically over recent time. Now they are successfully deployed as a part of many e-commerce sites, offering several important business benefits: increasing the number of items sold, selling more diverse items, increasing user satisfaction and loyalty, helping to understand what the user wants.

Collaborative filtering This technique leverages recommendations produced by a community of users to deliver recommendations to an *active user*—therefore it is sometimes referred as 'people-to-people correlation'. It is based on the fact that individuals often rely on recommendations provided by their peers making daily decisions and that similar people have similar tastes.

Content-based These type of recommender systems associate the derived content of the items with the user profile or characteristic. The system learns to recommend items similar to those that the user liked in the past, so the prediction is based on similarity between items. This approach is often used in Web recommenders and News filtering (where the content of the item is well-defined).

Knowledge-based These recommender systems rely on some kind of external knowledge about items. They are suitable for certain scenarios, where collaborative or content-based approaches show limitations. For example, buying a house, a car, or a computer is not a frequently made decision. Pure collaborative filtering system would not perform well, because of the low number of available ratings. For content-based systems the time span between ratings would make them useless. The main drawback of knowledge-based systems is a need for knowledge acquisition. It is also important to rank recommended items according to their utility for the user. Each item should be evaluated according to a predefined set of dimensions that provide an aggregate view.

3.4.3 Knowledge Based Approach for Recommendation

Approach based on action rule mining for developing a knowledge base of recommender systems presents a new way in machine learning. The concept of *action rules* was proposed by Ras and Wieczorkowska in 2000 [12] and is described in this chapter in a greater detail.

In [12] algorithm of rule extraction was applied to a sampling containing 20,000 tuples of a large banking database containing more that 10 million customers. In that case, action rules suggest how to shift bank's customers from the low-profit group into high-profit group—in particular which special offers should be made by a bank.

Action rule approach can be further enhanced by so called *meta-actions*, which help control the actions and act as a triggering mechanism for action rules. Meta actions are mechanism used to acquire knowledge about possible transitions in the information systems and their causes. The concept of meta-action was initially proposed in [13] to mine actionable patterns and then formally defined and used to discover action rules based on tree classifiers [14]. While extraction methods for action rules have been quite mature so far, the area of mining meta-actions is still developing. Work in [15] presents a methodology in medical field, but it concentrates on selecting meta-actions to achieve preferable effect given that meta-actions are already known.

In the proposed approach within this research, meta-actions are extracted from text and were used as a source of generating the triggers of action rules and ultimately used to improve Net Promoter Score ratings.

3.5 Text Analytics and Sentiment Analysis Tools

Edward Abbey once said: "Sentiment without action is the ruin of the soul". Indeed, it is important to mention some real-world applications of opinion mining and sentiment analysis. Growing demand for text analytics tools has raised the profile of specialized vendors such as Attensity OdinText [16], Clarabridge [17] and Kana—previously Overtone [18], which perform trended and basic root-cause analysis of customers' comments. SAS Institute, IBM SPSS, SAP (Insight) and Tibco (Insightful) offer tools for analyzing text for predictive insights. Lexalytics [19], Nstein [20] and Teragram—a division of SAS [21] offer text mining specialized for sentiment analysis. Some solutions attempt at recognizing importance of issues based on voice audio recordings and volume analysis—Verint Systems [22]. Rosetta Stone [23] is a solution using IBM SPSS text analytics software to analyze answers to open-ended questions in surveys of current and potential customers. It uses the resulting insights to drive decisions on advertising, marketing and product development, strategic planning as well as identify strengths and weaknesses of products.

Choice Hotels and Gaylord Hotels both applied text analytics software from Clarabridge [17] to quickly gather sentiment out of thousands of customer satisfaction surveys gathered each day. The software recognizes specific positive and negative comments and associates them with specific hotel locations, facilities, service, rooms and employee shifts. The feedback results in an immediate customer service response (through calls or letters) to acknowledge and apologize for problems. More important is that the system allows chain and facility managers track trends so that to spot problems and best practices.

Besides business, opinions matter a great deal in politics (example works: [24, 25]). Some work has focused on understanding what voters are thinking.

Sentiment-analysis and opinion-mining systems also have an important potential role as enabling technologies for other systems. One possibility is an augmentation to *recommendation systems*. Question answering is another area where sentiment can prove to be useful.

As one can see, sentiment-analysis technologies have many potential applications. In recent years sentiment analysis applications have spread to almost every possible domain, from consumer products, services, healthcare, and financial services to social events and political elections. As J.Ellen Foster said in 1893: "Sentiment is the mightiest force in civilization …".

References

1. Customer Satisfaction Surveys: Questions and Templates. https://www.surveymonkey.com/mp/csat/. Accessed: 2017-04-26.
2. Client Heartbeat: Customer Satisfaction Software Tool. https://www.clientheartbeat.com/. Accessed: 2017-04-26.
3. SurveyGizmo | Professional Online Survey Software and Tools. https://www.surveygizmo.com/. Accessed: 2017-04-26.
4. Customer Sure Customer Feedback Software | CustomerSure. http://www.customersure.com/. Accessed: 2017-04-26.
5. Online Survey and Benchmarking Application | Floq. http://floqapp.com/. Accessed: 2017-04-26.
6. Temper - Find out how your customers feel about every aspect of your business. https://www.temper.io/. Accessed: 2017-04-26.
7. The World's Leading Research and Insights Platform | Qualtrics. https://www.qualtrics.com/. Accessed: 2017-04-26.
8. A. Hausman, B. Noori, and M. Hossein Salimi. A decision-support system for business-to-business marketing. *Journal of Business & Industrial Marketing*, 20(4/5):226–236, 2005.
9. C. Rygielski, J.-C. Wang, and D. C. Yen. Data mining techniques for customer relationship management. *Technology in Society*, 24(4):483 – 502, 2002.
10. J. Bobadilla, F. Ortega, A. Hernando, and A. GutiéRrez. Recommender systems survey. *Know.-Based Syst.*, 46:109–132, July 2013.
11. X. Zhang and H. Wang. Study on recommender systems for business-to-business electronic commerce. *Communications of the IIMA*, 5(4):46–48, 2005.
12. Z. W. Ras and A. Wieczorkowska. *Action-Rules: How to Increase Profit of a Company*, pages 587–592. Springer Berlin Heidelberg, Berlin, Heidelberg, 2000.
13. K. Wang, Y. Jiang, and A. Tuzhilin. Mining actionable patterns by role models. In L. Liu, A. Reuter, K.-Y. Whang, and J. Zhang, editors, *ICDE*, page 16. IEEE Computer Society, 2006.

14. Z. W. Ras and A. Dardzinska. Action rules discovery based on tree classifiers and meta-actions. In *Foundations of Intelligent Systems, 18th International Symposium, ISMIS 2009, Prague, Czech Republic, September 14–17, 2009. Proceedings*, pages 66–75, 2009.
15. H. Touati, J. Kuang, A. Hajja, and Z. Ras. Personalized action rules for side effects object grouping. *International Journal of Intelligence Science*, 3(1A):24–33, 2013.
16. OdinText. http://odintext.com/. Accessed: 2017-04-26.
17. Choice Hotels deploys Clarabridge Text Mining Solution. http://www.clarabridge.com/press/choice-hotels-deploys-clarabridge-text-mining-solution/. Accessed: 2017-04-26.
18. Text Analytics Software I KANA. http://www.kana.com/text-analytics. Accessed: 2017-04-26.
19. Text Analytics Software SaaS and On-Premise I Lexalytics. https://www.lexalytics.com/. Accessed: 2017-04-26.
20. Semantic Navigation Search Appliance Application I OpenText. http://www.opentext.com/what-we-do/products/customer-experience-management/web-content-management/opentext-semantic-navigation. Accessed: 2017-04-26.
21. Teragram Linguistic Technologies I SAS. http://www.sas.com/enus/software/teragram.html. Accessed: 2017-04-26.
22. Voice of the customer - VOC Solutions I Verint Systems. http://www.verint.com/solutions/customer-engagement-optimization/voice-of-the-customer-analytics/. Accessed: 2017-04-26.
23. Rosetta stone speaks fluent customer satisfaction with text analytics software from spss inc. http://www.businesswire.com/news/home/20090615005156/en/Rosetta-Stone-Speaks-Fluent-Customer-Satisfaction-Text. Accessed: 2017-04-26.
24. M. Laver and J. Garry. Extracting Policy Positions from Political Texts Using Words as Data. *American Political Science Review*, 97:311–331, May 2003.
25. T. Mullen and R. Malouf. A preliminary investigation into sentiment analysis of informal political discourse. In *AAAI Symposium on Computational Approaches to Analysing Weblogs (AAAI-CAAW)*, pages 159–162, 2006.

Chapter 4
Background

This chapter presents theoretical background of the proposed approach: Knowledge Discovery and Decision Reduct techniques, Text Mining for Sentiment Analysis and Visualiztaion techniques for Data Analytics.

4.1 Knowledge Discovery

Addressing the problem of feature analysis, there are two approaches to feature evaluation and its importance towards classification problem.

The first one is based on the discrimination power of a set of features and how the classification problem (in terms of accuracy) is affected if one or more of them in a dataset are discarded. It always starts with the set of all features used in classification. This approach is logic-based and it can be called top-down approach.

The second one is a statistic-based and it can be called bottom-up approach. It talks about the discrimination power of a single feature or a small set of features. It does not make any reference to discrimination power of combined effect of features together. To compare them, one can say that the first approach is focused more on minimizing loss of knowledge, the second one more on maximizing knowledge gain.

4.1.1 Decision Reducts

To solve decision problems as stated in the previous chapter applying attribute reduction techniques is proposed. The one proposed is based on decision reducts and stems from rough set theory (logic-based).

Rough set theory is a mathematical tool for dealing with ambiguous and imprecise knowledge, which was presented by Polish mathematician Professor Pawlak in 1982

© Springer Nature Switzerland AG 2020
K. Tarnowska et al., *Recommender System for Improving Customer Loyalty*,
Studies in Big Data 55, https://doi.org/10.1007/978-3-030-13438-9_4

[1]. The rough set theory handles data analysis organized in the form of tables. The data may come from experts, measurements or tests. The main goal of the data analysis is a retrieval of interesting and novel patterns, associations, more precise problem analysis, as well as designing a tool for automatic data classification.

Attribute reduction is an important concept of rough set for data analysis. The main idea is to obtain decisions or classifications of problems on the conditions of maintaining the classification ability of the knowledge base. The basic concepts of the theory are introduced in the subsections below.

4.1.1.1   Information Systems

A concept of *Information System* stems from the theory of rough sets, developed by Zdzislaw Pawlak at the beginning of 1980s.

Definition 1 An *Information System* is defined as a pair $S = (U, A)$, where U is a nonempty, finite set, called *the universe*, and A is a nonempty, finite set of attributes i.e. $a : U \rightarrow V_a$ for $a \in A$, where V_a is called the domain of a [2].

Elements of U are called *objects*. A special case of *Information Systems* is called a *Decision Table* [3].

4.1.1.2   Decision Tables

In a decision table, some attributes are called *conditions* and the others are called *decisions*. In many practical applications, decision is a singleton set. For example, in table in Fig. 4.2 decision is an attribute specifying *Promoter Status*. The conditions would be all the attributes that determine *Promoter Status*, that is, question benchmarks and also other attributes (geographical, temporal, etc.).

Based on knowledge represented in a form of a decision table, it is possible to model and simulate decision-making processes. The knowledge in a decision table is represented by associating or identifying decision values with some values of conditional attributes.

For extracting action rules, it is also relevant to differentiate between so-called *flexible* attributes, which can be changed, and *stable* attributes [2], which cannot be changed. $A = A_{St} \cup A_{Fl}$, where A_{St} and A_{Fl} denote *stable* attributes and *flexible* attributes respectively. Example of stable attributes in customer data would be client's and survey's characteristics, while flexible would be assessment areas (benchmarks), which can be changed by undertaking certain actions (for example, staff training).

4.1.1.3   Reducts

In decision systems not every attribute in the database is necessary for the decision-making process. The goal is to choose some subset of attributes essential for this. It leads to the definition of *reducts*, that is, minimal subsets of attributes that keep the characteristics of the full dataset. In the context of action rule discovery an *action reduct* is a minimal set of attribute values distinguishing a favorable object from another. In the considered application area, it is of interest to find unique characteristics of the satisfied customers that can be used by the company to improve the customer satisfaction of 'Detractors'. There is a need to find a set of distinct values or unique patterns from the 'Promoter' group that does not exist in the 'Detractor' group in order to propose an actionable knowledge. Some of attribute values describing the customers can be controlled or changed, which is defined as an action. Before defining formally a *reduct*, it is necessary to introduce a *discernibility relation*.

Definition 2 Let objects $x, y \in U$ and set of attributes $B \subset A$. We say that x, y are *discernible* by B when there exists $a \in B$ such that $a(x) \neq b(y)$. x, y are *indiscernible* by B when they are identical on B, that is, $a(x) = b(y)$ for each $a \in B$. $[x]_B$ denotes a set of objects indiscernible with x by B.

Furthermore, following statements are true:

- for each objects x, y either $[x]_B = [y]_B$ or $[x]_B \cap [y]_B = \emptyset$,
- *indiscernibility relation* is an equivalence relation,
- each set of attributes $A \subset B$ determines a partition of a set of objects into disjoint subsets.

Definition 3 A set of attributes $B \subset A$ is called *reduct of the decision table* if and only if:

- B keeps the discernibility of A, that is, for each $x, y \in U$, if x, y are discernible by A, then they are also discernible by B,
- B is irreducible, that is, none of its proper subset keeps discernibility properties of A (that is, B is minimal in terms of discernibility).

The set of attributes appearing in every reduct of information system A (decision table DT) is called *the core*.

For example, in order to allow for local and temporal analysis of the considered application, the yearly global NPS data (2011–2015) is divided into separate client data (38 datasets for each year in total). For each such extracted dataset a corresponding feature selection, transformation is performed and then an attribute reduction algorithm is run (with RSES package [4]).

4.1.2 Classification

To guarantee mining high quality action rules, the best classifiers have to be constructed first. Also, the results of classification provide an overview of the consistency of knowledge hidden in the dataset. The better the classification results the more consistent and accurate knowledge stored in the data. Also, better ability of the system to recognize Promoters/Passives/Detractor correctly is a foundation for the system to give accurate results.

For example, to track the accuracy of the models built on the yearly client data classification experiments for each client's dataset and for each year has been performed. Evaluation was performed with 10-fold cross-validation on decomposition tree classifiers with RSES package (Rough Set Exploration System [4]).

Decomposition trees are used to split data set into fragments not larger than a predefined size. These fragments, after decomposition represented as leaves in decomposition tree, are supposed to be more uniform. The subsets of data in the leaves of decomposition tree are used for calculation of decision rules.

The results from each classification task: accuracy, coverage, confusion matrix were saved.

4.1.2.1 Decision Rule

The decision rule, for a given decision table, is a rule in the form: $(\phi \rightarrow \delta)$, where ϕ is called *antecedent* (or *assumption*) and δ is called *descendant* (or *thesis*) of the rule. The antecedent for an atomic rule can be a single term or a conjunction of k elementary conditions: $\phi = p_1 \wedge p_2 \wedge \cdots \wedge p_n$, and δ is a decision attribute. Decision rule describing a class K_j means that objects, which satisfy (match) the rule's antecedent, belong to K_j.

In the context of prediction problem, decision rules generated from training dataset, are used for classifying new objects (for example classifying a new customer for NPS category). New objects are understood as objects that were not used for the rules induction (new customers surveyed). The new objects are described by attribute values (for instance a customer with survey's responses). The goal of classification is to assign a new object to one of the decision classes.

4.1.3 Action Rules

Action rule concept was firstly proposed by Ras and Wieczorkowska in [2], and since then they have been successfully applied in many domain areas including business [2], medical diagnosis and treatment [5], music automatic indexing and retrieval [6].

Action rules present a new way in machine learning domain that solve problems that traditional methods, such as classification or association rules cannot handle. The purpose is to analyze data to improve understanding of it and seek specific actions (recommendations) to enhance the decision-making process. An *action* is understood as a way of controlling or changing some of attribute values in an information system to achieve desired results [7]. An *action rule* is defined [2] as a rule extracted from an information system, that describes a transition that may occur within objects from one state to another, with respect to decision attribute, as defined by the user. Decision attribute is a distinguished attribute [2], while the rest of the attributes are partitioned into stable and flexible attributes.

In nomenclature, action rule is defined as a term: $[(\omega) \wedge (\alpha \rightarrow \beta) \rightarrow (\Phi \rightarrow \Psi)]$, where ω denotes conjunction of fixed stable attributes, $(\alpha \rightarrow \beta)$ are proposed changes in values of flexible attributes, and $(\Phi \rightarrow \Psi)$ is a desired change of decision attribute (action effect).

So, in the considered domain, decision attribute is *PromoterStatus* (with values *Promoter, Passive, Detractor*). Let us assume that Φ means 'Detractors' and Ψ means 'Promoters'. The discovered knowledge would indicate how the values of flexible attributes need to be changed under the condition specified by stable attributes so the customers classified as Detractors should become Promoters. So, an action rule discovery applied to customer data would suggest a change in flexible attribute values, such as different benchmarks to help "reclassify" or "transit" an object (customer) to a different category (Passive or Promoter) and consequently, attain better overall customer satisfaction.

An action rule is built from *atomic action sets*.

Definition 4 *Atomic action term* is an expression $(a, a_1 \rightarrow a_2)$, where a is attribute, and $a_1, a_2 \in V_a$, where V_a is a domain of attribute a.

If $a_1 = a_2$ then a is called stable on a_1.

Definition 5 By *action sets* we mean the smallest collection of sets such that:

1. If t is an atomic action term, then t is an action set.
2. If t_1, t_2 are action sets, then $t_1 \wedge t_2$ is a candidate action set.
3. If t is a candidate action set and for any two atomic actions $(a, a_1 \rightarrow a_2)$, $(b, b_1 \rightarrow b_2)$ contained in t we have $a \neq b$, then t is an action set. Here b is another attribute $(b \in A)$, and $b_1, b_2 \in V_b$.

Definition 6 By an *action rule* we mean any expression $r = [t_1 \Rightarrow t_2]$, where t_1 and t_2 are action sets.

The interpretation of the action rule r is, that by applying the action set t_1, we would get, as a result, the changes of states in action set t_2. So, action rule suggests the smallest set of necessary actions needed for switching from current state to another within the states of the decision attribute.

The ultimate goal of building an efficient recommender system is to provide actionable suggestions for improving a client's performance (improving its NPS efficiency rating). Extracting action rules is one of the most operative methods here and it has been applied to various application areas like medicine (developing medical treatment methods) or sound processing.

The application of action rules in the considered domain: clients' datasets are mined for finding action rules, that is, rules which indicate actions to be taken in order to increase Net Promoter Score. The results of action rule mining are used in the process of generating recommendations in the proposed knowledge-based system.

4.1.4 Clustering

It is believed that clients can collaborate with each other by exchanging knowledge hidden in datasets and they can benefit from others whose hidden knowledge is similar. In order to recommend items (actions to improve in the service, products), one needs to consider not only historical feedback of customers for this client, but also look at clients who are similar in some way, but perform better. The concept of semantic similarity is used to compare clients, which is defined as similarity of their knowledge concerning the meaning of three concepts: Promoter, Passive, and Detractor. Clients who are semantically close to each other can have their datasets merged and the same considered as a single client from the business perspective (customers have similar opinion about them).

In the proposed approach hierarchical clustering algorithm is used to generate the nearest neighbors of each client. Given the definition of semantic similarity, the distance between any pairs of clients are quantified in a semantic way and the smaller the distance is, the more similar the clients are. A semantic similarity-based distance matrix is built on top of the definition. With the distance matrix, a hierarchical clustering structure (dendrogram) is generated by applying an agglomerative clustering algorithm (with the R package).

A dendrogram is a node-link diagram that places leaf nodes of the tree at the same depth. If describing it using tree-structure-based terminology, every leaf node in the dendrogram represents the corresponding client as the number shown, and the depth of one node is the length of the path from it to the root, so the lower difference of the depth between two leaf nodes, the more semantically similar they are to each other. With the dendrogram it is easy to find out the groups of clients which are relatively closer to each other in semantic similarity.

4.2 Text Mining

This subsection describes the background necessary to understand approach and techniques used in the built Recommender System to mine text data of customer reviews. It also surveys the existing approaches and applications in the area of text mining with a focus on sentiment analysis.

4.2.1 Sentiment Analysis

Sentiment analysis and opinion mining is the field of study that analyzes people's opinions, sentiments, evaluations, attitudes, and emotions from written language. It became one of the most active research areas in natural language processing, widely studied in data mining, Web mining and text mining. Sentiment analysis systems are being applied in almost every business and social domain because opinions are central to almost all human activities and are key influencers of our behaviors.

Sentiment analysis is usually part of a larger text analytics framework and involves analyzing subjective part for their polarity (whether they denote positive opinion or negative opinion). Analyzing the opinion can be performed at three levels [8]: extracting the overall sentiment of an entire comment (document-level), on each sentence of a comment or in reference to certain aspects or features of the product/service (e.g., price, staff, service). Document-level sentiment analysis is mostly based on supervised learning techniques (classification), but there are also some unsupervised methods [9].

Sentiment classification can be obviously formulated as a classification problem with two decision classes: positive and negative. Existing supervised methods can be readily applied to sentiment classification: Naive Bayes, Support Vector Machines (SVM), etc. The earliest work of automatic sentiment classification at document level is [10] where such approach was taken to classify movie reviews from IMDB into two classes, positive and negative. Features used to classify could be: terms and their frequency, part of speech tags, opinion words, syntactic dependency, negation [11]. Besides binary prediction of sentiment, there was also research conducted aimed at predicting the rating scores of reviews [12]. In this case problem was formulated as a regression problem (as the rating scores are ordinal), and solved using SVM regression. There were also unsupervised methods proposed for sentiment classification: Turney et al. [13] performed classification based on some fixed syntactic patterns that are likely to be used to express opinions. The syntactic patterns were composed of part-of-speech tags.

At the sentence level, each sentence in the document is analyzed and classified as either positive or negative. The methods are similar as in case of document-level sentiment analysis. In [14] sentence level sentiment analysis is using rules based on clauses of a sentence.

Sentiment classification does not try to find concrete features that were commented on—therefore, its granularity of analysis is different to that of aspect-based sentiment analysis.

4.2.2 Aspect-Based Sentiment Analysis

Opinions extracted at the document or at the sentence level often do not provide the necessary detail needed for some applications which require sentiment analysis on some aspects or features of the object (on what people exactly liked and did not like). Aspect-level sentiment analysis performs a fine-grained analysis. It is based on the idea that an opinion consists of a *sentiment* (positive or negative) and a *target* (of opinion). It helps to understand the sentiment problem better and to address mixed opinions, such as: "Although the service is not that great, I still love this restaurant". This sentence has a positive tone, but in the aspect of *service* is negative.

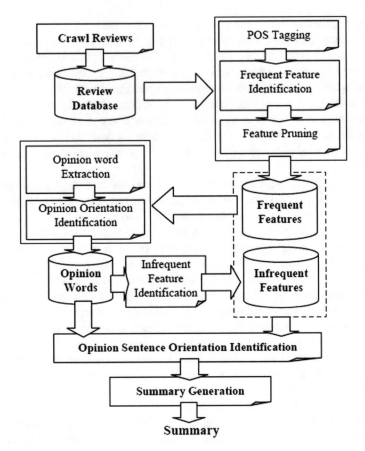

Fig. 4.1 Feature-based opinion summarization process. *Source* [15]

The major tasks in the aspect-based sentiment analysis are (Damerau, 2010):

- Aspect extraction (feature identification).
- Recognition of polarity towards given aspect (positive/negative/neutral).
- Producing structured summary of opinions about aspects, which can be further used for qualitative and quantitative analyses.

The text mining process for Web reviews involving aspect-based sentiment analysis and summarization is described in paper [15], which is considered pioneer work on feature-based opinion summarization. Three subtasks of generating feature-based summaries are defined: (1) identifying features of the product, (2) identifying review opinionated sentences, (3) producing summaries. The framework for opinion summarization system is depicted in Fig. 4.1.

The system crawls the Web for the customer reviews and stores them in a database. Then it extracts most frequent features on which people expressed their opinion (with the use of part-of-speech tagging). They use association rule mining based on the Apriori algorithm to extract frequent itemsets as explicit product features. Itemsets that have support at least equal to minimum support are called frequent itemsets [16]. Secondly, opinion words are extracted using the resulting frequent features and semantic orientations of the opinion words are determined based on WordNet and positive/negative word dictionary. This way opinion sentences are identified— opinion sentence must contain one or more feature words as well as opinion words describing these features. In the last steps, the orientation of each opinion sentence is identified and a final summary is produced. This approach also handles negations and 'but-clauses'. At the end, it applies an opinion aggregation function to determine the final orientation of the opinion on each object feature in the sentence.

4.2.3 Aspect Extraction

In case of lack of domain knowledge and consequently domain-specific aspect dictionary an important step in sentiment analysis is aspect extraction. Aspect extraction can be seen as an information extraction task. There are four main approaches:

- Extraction based on frequent nouns and noun phrases.
- Extraction by exploiting opinion and target relations (syntactical relations).
- Supervised learning.
- Topic modeling/unsupervised learning.

There exist variety of methods for aspect extraction, such as word n-grams, bi-grams, word cluster, casting, POS tagging, parse dependencies, relations and punctuations marks. Supervised learning techniques include: Hidden Markov Models (HMM) and Conditional Random Fields (CRF).

Topic modeling is an unsupervised learning method that assumes each document consists of a mixture of topics and each topic is a probability distribution over words. There were two basic models: pLSA (Probabilistic Latent Semantic Analysis—[17]) and LDA (Latent Dirichlet allocation—[18]).

In paper [19] an unsupervised information extraction system called OPINE was developed. OPINE first extracts noun phrases from reviews and retains those with frequency greater than an experimentally set threshold. Yi et al. [20] proposed a term extraction technique based on heuristics and selection algorithms. They are also using sentiment extraction pattern as a processing step to opinion extraction. Zhuang et al. [21] proposed a multi-knowledge based approach for movie review and summarization—they used the keyword list and dependency relation templates together to mine explicit feature-opinion pairs. In [22] method based on syntactical dependency relations was presented for extracting the product feature and identifying opinions that associate with product features in each sentence. Their approach is presented in Fig. 4.2. First, they perform pre-processing—parsing and dependency analysis. The reviews are parsed by the use of the Stanford Parser [23]. As a result a dependency tree is generated (Fig. 4.3). While parsing the sentence, noun phrases

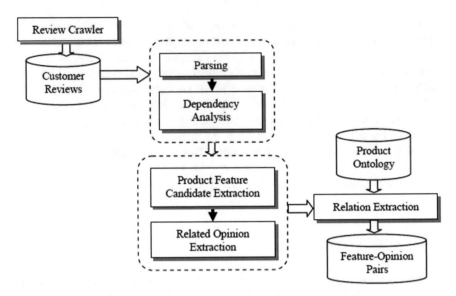

Fig. 4.2 Architecture of the system for mining Feature-Opinion words based on syntactical dependency. *Source* [22]

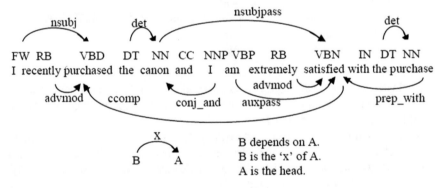

Fig. 4.3 Example of syntactical dependency tree. *Source* [22]

are identified as a product feature candidates using linguistic pattern. Then for each
product feature candidate in every dependency parse tree, related opinion words are
searched for amongst adjectives and verbs. A set of candidate feature-opinion pairs is
generated and then probabilistic based model (Maximum Entropy Model) is used to
predict the relevance of each such relationship. Additionally, authors of these papers
proposed using product ontology to resolve the problem of incompatible terminology
(different customers referring to the same product features using different terminology). Ontology contains encoded semantic information and provides a source of
shared and precisely defined terms.

After aspect extraction, optional step is to group these into synonymous aspect categories, so that each category represents a unique aspect. For example, "call quality" and "voice quality" refer to the same aspect for phones. Carenini et al. [24] proposed the first method to deal with this problem. It was based on several similarity metrics defined using string similarity, synonyms, and lexical distances measured using WordNet. In Yu et al. [25] a more sophisticated method was proposed to also use publicly available hierarchies/taxonomies of products and the actual product reviews to generate the ultimate aspect hierarchies. In Zhai et al. [26], a semi-supervised learning method was proposed to group aspect expressions into some user-specified aspect categories.

4.2.4 Polarity Calculation

The most important indicators of sentiments are *sentiment words* (*opinion words*). A list of such words is called *sentiment lexicon* (*opinion lexicons*). Opinion lexicons are resources that associate sentiment orientation and words. Over the years, researchers have designed numerous algorithms to compile such lexicons. These lexicons can be used not only for polarity detection but also for further supervised expansions of lexicons, as presented in [27] (for tweets).

4.2.4.1 Bing Liu

Hu and Liu [15] used a sentiment orientation labeled list of adjectives (this list was expanded with some nouns by Liu et al. [28]). It consists of two lists: one has positive entries (in number of 2003) and the other—negative (4782).

4.2.4.2 SentiWordNet

The most popular sentiment dictionary is SentiWordNet, built on top of WordNet (a lexical database for the English language—[29]), where a pair of positive and negative polarity score is assigned to each sense of a word [30]. SentiWordNet entry for each word comprises of all the possible parts of speech in which the word could appear, all the senses corresponding to each part of speech and a pair of polarity scores associated with each sense. There are 28,431 sentiment bearing entries (out of total 86,994 WordNet terms). The range for positive and negative polarity scores is from 0 to 1. The default algorithm (provided by SentiWordnet website) calculates an overall polarity as: Positive score—Negative Score, for each sense of a word. Next it calculates a weighted sum of all the overall polarities for all the senses of the word, with the weights defined as the ranks of senses. The polarity scores in SentiWordNet were generated automatically using a semi-supervised method described in [31].

4.2.4.3   AFINN

AFINN is a strength-oriented lexicon [32] with positive words (564 in total) scored from 1 to 5 and negative words (964) scored from -1 to -5. It also includes slang, obscene words, acronyms and Web jargon.

4.2.4.4   MPQA Subjectivity Lexicon

This lexicon was created by Wilson et al. [33] as a part of their system Opinion-Finder. The lexicon consists of positive (2,295), negative (4,148) and neutral words (424).

4.2.4.5   NRC Emotion Lexicon

This is emotion-oriented lexicon created by conducting a tagging process on the crowdsourcing Amazon Mechanical Turk platform [34]. The words are annotated by eight emotions: joy, trust, sadness, anger, surprise, fear, anticipation, disgust, as well as two polarity classes: positive (2,312) and negative (3,324). There are also words not associated with any emotional state and tagged as neutral (7,714).

4.2.5   Natural Language Processing Issues

Although sentiment words are important for sentiment analysis, only using them is not sufficient. Natural Language recognition is much more complex:

- Words can have opposite orientations in different application domain.
- Sentiment word might not express opinion in question (interrogative) sentences and conditional sentences (e.g. "Can you tell me which Sony camera is *good*?, "If I can find a *good* camera in the shop, I will buy it").
- Sarcastic sentences are hard to deal with.
- Sentences without sentiment words can also imply opinions (e.g. "This dishwasher uses a lot of water").

Besides, sentiment analysis as each NLP task must handle coreference, negation, disambiguation, comparative sentences.

4.2.6   Summary Generation

In most sentiment analysis application, one needs to study opinions from many people due to subjective nature of opinions. Some form of summary is needed.

Fig. 4.4 Example of summary generation. *Source* [15]

Feature: **picture**

Positive: 12

- Overall this is a good camera with a really good picture clarity.
- The pictures are absolutely amazing - the camera captures the minutest of details.
- After nearly 800 pictures I have found that this camera takes incredible pictures.

...

Negative: 2

- The pictures come out hazy if your hands shake even for a moment during the entire process of taking a picture.
- Focusing on a display rack about 20 feet away in a brightly lit room during day time, pictures produced by this camera were blurry and in a shade of orange.

It is a next step, after detecting aspect, opinion words and calculating polarity. It involves two basic steps as described in [15], pioneering work on review mining and summarization:

- For each feature, associated opinion sentences are put into positive and negative buckets (according to the calculated polarity). Optionally, a count can be calculated so that to show how many there are positive/negative comments about a particular feature.
- Features are ranked according to the frequency of their appearances in the reviews.

Example of reviews' summary is shown in Fig. 4.4—for the feature *"picture"* of a digital camera:

4.2.7 Visualizations

The work in "Visualizing Multiple Variables Across Scale and Geography" [35] from the 2015 IEEE VIS conference attempted a multidimensional attribute analysis varying across time and geography. Especially interesting is approach for analyzing attributes in terms of their correlation to the decision attribute, which involves both global and local statistics. The sequence of panels allows for a more fine-grained, sequential analysis by discovering strongly correlated variables at a global level, and then investigating it through geographical variation at the local level (see Fig. 4.5). The approach presented in the paper supports a correlation analysis in many dimensions, including geography and time, as well as, in our case—particular company, which bears similarity to the problem within this research. The visualization proposed in the paper helps in finding more discriminative profiles when creating geo-demographic classifiers. For review summarization purposes, often a variety of visualization methods are deployed. In its very basic form, for example, Amazon

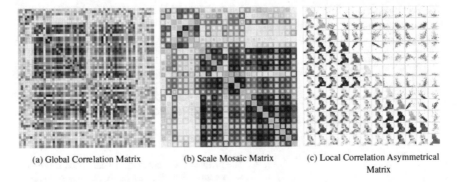

(a) Global Correlation Matrix   (b) Scale Mosaic Matrix   (c) Local Correlation Asymmetrical
                                                                                                Matrix

Fig. 4.5 Multivariate comparison across scale and geography showing correlation in [35]

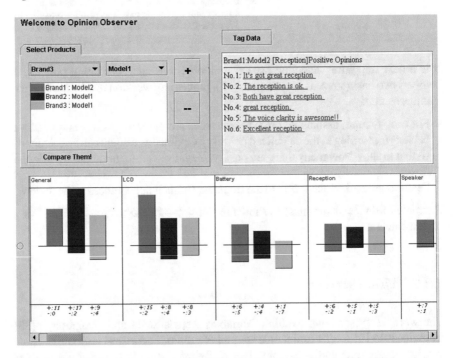

Fig. 4.6 Visualization for aspect-based review summarization developed within [28]

displays an average rating and a number of reviews next to it. Mousing over the
stars brings up a histogram of reviewer ratings annotated with counts for the 5-star
reviews, 4-star reviews, etc.

As another example, a sample output of the Opinion Observer system [28] is
depicted in Fig. 4.6, where the portion of a bar projecting above the centered "horizon"
line represents the number of positive opinions about a certain product feature, and
the portion of the bar below the line represents the number of negative opinions.

Fig. 4.7 Rose plots technique for visualizing an affective content of the documents. *Source* [36]

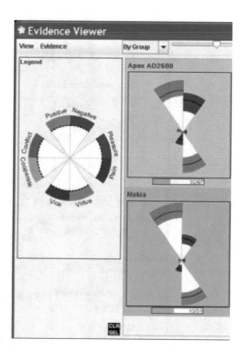

While the data for the features are presented sequentially in Fig. 4.6 (first "General", then "LCD", and so forth), an alternative visualization technique called a *rose plot* is exemplified in Fig. 4.7, which depicts a sample output of the system developed by Gregory et al. [36].

The median and quartiles across a document sub-collection of the percentage of positive and negative words per document, together with similar data for other possible affect-classification dimensions, are represented via a variant of box plots.

Morinaga et al. [37] proposed to visualize degrees of association between products and opinion-indicative terms of a pre-specified polarity. Principal component analysis is applied to two-dimensional visualization (Fig. 4.8), such that nearness corresponds to strength of association.

4.2.8 Measuring the Economic Impact of Sentiment

Reviews influence both purchasing decisions of other customers reading the reviews as well as product manufacturers making product-development, marketing and advertising decisions. However subjective perception of "the influenced" and reality might differ. Therefore, a key element is to understand the real economic impact of sentiment expressed in surveys and reviews. The results of such analysis can be used by

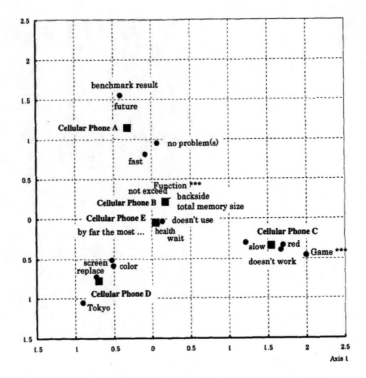

Fig. 4.8 Principal-Components-analysis visualization of associations between products (squares) and automatically selected opinion-oriented terms (circles). *Source* [37]

companies to estimate how much effort and resources should be allocated to address the issues.

There already have been conducted a vast study within the economics and market-ing literature to find out whether the polarity has a measurable, significant influence on customers (e.g. [38, 39]).

The most popular approach is to use hedonic regression to analyze the value and the significance of different item features to some function, such as a measure of utility to the customer, using historic data [40]. Specific economic functions under examination include revenue, revenue growth, stock trading volume, etc. It is worth noticing that different subsegments of the consumer population may react differently. Additionally, in some studies, positive ratings have an effect but negatives ones do not, and in other studies the opposite effect is seen. Anyway, in most studies a positive correlation affect is observed between survey polarity and economic effect—correlation statistically significant (e.g. [41–47]).

Work in [48] attempts to assign a "dollar value" to various adjective-noun pairs, adverb-verb pairs and similar lexical configurations.

References

1. Z. Pawlak and W. Marek. Rough sets and information systems, 1981.
2. Z. W. Ras and A. Wieczorkowska. *Action-Rules: How to Increase Profit of a Company*, pages 587–592. Springer Berlin Heidelberg, Berlin, Heidelberg, 2000.
3. Z. Pawlak. Rough sets and decision tables, 1985.
4. RSES 2.2 User's Guide. http://logic.mimuw.edu.pl/~rses.
5. H. Wasyluk, Z. Ras, and E. Wyrzykowska. Application of action rules to hepar clinical decision support system. *Experimental and Clinical Hepatology*, 4(2):46–48, 2008.
6. Z. W. Ras and A. Dardzinska. From data to classification rules and actions. *Int. J. Intell. Syst.* 26(6):572–590, 2011.
7. S. Im, Z. W. Ras, and L. Tsay. Action reducts. In *Foundations of Intelligent Systems - 19th International Symposium, ISMIS 2011, Warsaw, Poland, June 28–30, 2011. Proceedings*, pages 62–69, 2011.
8. R. Feldman. Techniques and applications for sentiment analysis. *Commun. ACM*, 56(4):82–89, Apr. 2013.
9. N. Indurkhya and F. J. Damerau. *Handbook of Natural Language Processing*. Chapman & Hall/CRC, 2nd edition, 2010.
10. B. Pang, L. Lee, and S. Vaithyanathan. Thumbs up?: Sentiment classification using machine learning techniques. In *Proceedings of the ACL-02 Conference on Empirical Methods in Natural Language Processing - Volume 10*, EMNLP '02, pages 79–86, Stroudsburg, PA, USA, 2002. Association for Computational Linguistics.
11. B. Pang and L. Lee. Opinion mining and sentiment analysis. *Found. Trends Inf. Retr.*, 2(1–2):1–135, Jan. 2008.
12. B. Pang and L. Lee. Seeing stars: Exploiting class relationships for sentiment categorization with respect to rating scales. *CoRR*, arXiv:abs/cs/0506075, 2005.
13. P. D. Turney. Thumbs up or thumbs down?: Semantic orientation applied to unsupervised classification of reviews. In *Proceedings of the 40th Annual Meeting on Association for Computational Linguistics*, ACL '02, pages 417–424, Stroudsburg, PA, USA, 2002. Association for Computational Linguistics.
14. A. Meena and T. V. Prabhakar. Sentence level sentiment analysis in the presence of conjuncts using linguistic analysis. In *Advances in Information Retrieval, 29th European Conference on IR Research, ECIR 2007, Rome, Italy, April 2–5, 2007, Proceedings*, pages 573–580, 2007.
15. M. Hu and B. Liu. Mining and summarizing customer reviews. In *Proceedings of the Tenth ACM SIGKDD International Conference on Knowledge Discovery and Data Mining*, KDD '04, pages 168–177, New York, NY, USA, 2004. ACM.
16. O. Daly and D. Taniar. *Exception Rules Mining Based on Negative Association Rules*, pages 543–552. Springer Berlin Heidelberg, Berlin, Heidelberg, 2004.
17. T. Hofmann. Probabilistic latent semantic indexing. In *Proceedings of the 22Nd Annual International ACM SIGIR Conference on Research and Development in Information Retrieval*, SIGIR '99, pages 50–57, New York, NY, USA, 1999. ACM.
18. D. M. Blei, A. Y. Ng, and M. I. Jordan. Latent dirichlet allocation. *J. Mach. Learn. Res.* 3:993–1022, Mar. 2003.
19. A.-M. Popescu and O. Etzioni. Extracting product features and opinions from reviews. In *Proceedings of the Conference on Human Language Technology and Empirical Methods in Natural Language Processing*, HLT '05, pages 339–346, Stroudsburg, PA, USA, 2005. Association for Computational Linguistics.
20. J. Yi and W. Niblack. Sentiment mining in webfountain. In K. Aberer, M. J. Franklin, and S. Nishio, editors, *ICDE*, pages 1073–1083. IEEE Computer Society, 2005.
21. L. Zhuang, F. Jing, and X.-Y. Zhu. Movie review mining and summarization. In *Proceedings of the 15th ACM International Conference on Information and Knowledge Management*, CIKM '06, pages 43–50, New York, NY, USA, 2006. ACM.

22. G. Somprasertsri and P. Lalitrojwong. Mining feature-opinion in online customer reviews for opinion summarization. *j-jucs*, 16(6):938–955, mar 2010. http://www.jucs.org/jucs166/miningfeatureopinionin.

23. M.-C. D. Marneffe and C. D. Manning. Stanford typed dependencies manual, 2008.

24. G. Carenini, R. T. Ng, and E. Zwart. Extracting knowledge from evaluative text. In *Proceedings of the 3rd International Conference on Knowledge Capture*, K-CAP '05, pages 11–18, New York, NY, USA, 2005. ACM.

25. J. Yu, Z.-J. Zha, M. Wang, and T.-S. Chua. Aspect ranking: Identifying important product aspects from online consumer reviews. In *Proceedings of the 49th Annual Meeting of the Association for Computational Linguistics: Human Language Technologies - Volume 1*, HLT '11, pages 1496–1505, Stroudsburg, PA, USA, 2011. Association for Computational Linguistics.

26. Z. Zhai, B. Liu, H. Xu, and P. Jia. Grouping product features using semi-supervised learning with soft-constraints. In *Proceedings of the 23rd International Conference on Computational Linguistics*, COLING '10, pages 1272–1280, Stroudsburg, PA, USA, 2010. Association for Computational Linguistics.

27. F. Bravo-Marquez, E. Frank, and B. Pfahringer. Positive, negative, or neutral: Learning an expanded opinion lexicon from emoticon-annotated tweets. In *Proceedings of the Twenty-Fourth International Joint Conference on Artificial Intelligence, IJCAI 2015, Buenos Aires, Argentina, July 25–31, 2015*, pages 1229–1235, 2015.

28. B. Liu, M. Hu, and J. Cheng. Opinion observer: Analyzing and comparing opinions on the web. In *Proceedings of the 14th International Conference on World Wide Web*, WWW '05, pages 342–351, New York, NY, USA, 2005. ACM.

29. C. Fellbaum, editor. *WordNet: an electronic lexical database*. MIT Press, 1998.

30. S. Baccianella, A. Esuli, and F. Sebastiani. Sentiwordnet 3.0: An enhanced lexical resource for sentiment analysis and opinion mining. In N. C. C. Chair), K. Choukri, B. Maegaard, J. Mariani, J. Odijk, S. Piperidis, M. Rosner, and D. Tapias, editors, *Proceedings of the Seventh International Conference on Language Resources and Evaluation (LREC'10)*, Valletta, Malta, may 2010. European Language Resources Association (ELRA).

31. A. Esuli and F. Sebastiani. Determining term subjectivity and term orientation for opinion mining. In *Proceedings of the European Chapter of the Association for Computational Linguistics (EACL)*, 2006.

32. F. A. Nielsen. A new anew: Evaluation of a word list for sentiment analysis in microblogs. *CoRR*, arXiv:abs/1103.2903, 2011.

33. T. Wilson, J. Wiebe, and P. Hoffmann. Recognizing contextual polarity in phrase-level sentiment analysis. In *Proceedings of the Conference on Human Language Technology and Empirical Methods in Natural Language Processing*, HLT '05, pages 347–354, Stroudsburg, PA, USA, 2005. Association for Computational Linguistics.

34. S. M. Mohammad and P. D. Turney. Crowdsourcing a word-emotion association lexicon. 29(3):436–465, 2013.

35. S. Goodwin, J. Dykes, A. Slingsby, and C. Turkay. Visualizing multiple variables across scale and geography. *IEEE Transactions on Visualization and Computer Graphics*, 22(1):599–608, Jan 2016.

36. M. L. Gregory, N. Chinchor, P. Whitney, R. Carter, E. Hetzler, and A. Turner. User-directed sentiment analysis: Visualizing the affective content of documents. In *Proceedings of the Workshop on Sentiment and Subjectivity in Text*, SST '06, pages 23–30, Stroudsburg, PA, USA, 2006. Association for Computational Linguistics.

37. S. Morinaga, K. Yamanishi, K. Tateishi, and T. Fukushima. Mining product reputations on the web. In *Proceedings of the Eighth ACM SIGKDD International Conference on Knowledge Discovery and Data Mining*, KDD '02, pages 341–349, New York, NY, USA, 2002. ACM.

38. C. Shapiro. Consumer information, product quality, and seller reputation. *Bell Journal of Economics*, 13(1):20–35, 1982.

39. C. Shapiro. Premiums for high quality products as returns to reputations. *The Quarterly Journal of Economics*, 98(4):659, 1983.

40. S. Rosen. Hedonic prices and implicit markets: Product differentiation in pure competition. *Journal of Political Economy*, 82(1):34–55, 1974.

41. N. Archak, A. Ghose, and P. G. Ipeirotis. Show me the money!: deriving the pricing power of product features by mining consumer reviews. In *KDD '07: Proceedings of the 13th ACM SIGKDD international conference on Knowledge discovery and data mining*, pages 56–65, New York, NY, USA, 2007. ACM.

42. S. Basuroy, S. Chatterjee, and A. S. Ravid. How Critical Are Critical Reviews? The Box Office Effects of Film Critics, Star Power, and Budgets. *Journal of Marketing*, 67(4):103–117, Oct. 2003.

43. J. A. Chevalier and D. Mayzlin. The effect of word of mouth on sales: Online book reviews. *Journal of marketing research*, 43(3):345–354, 2006.

44. C. Dellarocas, X. M. Zhang, and N. F. Awad. Exploring the value of online product ratings in revenue forecasting: The case of motion pictures. *Journal of Interactive Marketing*, 21(4):23–45, 2007.

45. Y. Liu, X. Huang, A. An, and X. Yu. Arsa: A sentiment-aware model for predicting sales performance using blogs. In *Proceedings of the 30th Annual International ACM SIGIR Conference on Research and Development in Information Retrieval*, SIGIR '07, pages 607–614, New York, NY, USA, 2007. ACM.

46. G. Mishne and N. Glance. Predicting movie sales from blogger sentiment. In *AAAI Symposium on Computational Approaches to Analysing Weblogs (AAAI-CAAW)*, pages 155–158, 2006.

47. R. Tumarkin and R. F. Whitelaw. News or noise? Internet postings and stock prices. *Financial Analysts Journal*, 57(3):41–51, May/June 2001.

48. A. Ghose, P. Ipeirotis, and A. Sundararajan. Opinion mining using econometrics: A case study on reputation systems. In *Proceedings of the 45th Annual Meeting of the Association of Computational Linguistics*, pages 416–423, Prague, Czech Republic, June 2007. Association for Computational Linguistics.

Chapter 5
Overview of Recommender System Engine

This chapter presents the architecture and mechanism ("engine") of the proposed Recommender System built from structured and unstructured customer data.

5.1 High-Level Architecture

The built recommender system is data-driven, that is, works based on the data collected historically year by year. The data has to be further preprocessed to be useful for the purpose of knowledge extraction. New features are added and the column values are standardized and completed. The dataset is divided into single-clients datasets. All these processes are described in the next section.

GUI module of the recommender system (in Java) implements main use cases the user interacts with the system (see Fig. 5.1). The user chooses (in an Overview panel) the year and survey category on what basis the analysis will be conducted. Next, the chosen client's dataset is expanded with other semantically similar (but better performing, in terms of NPS) clients' datasets (in HAMIS process tab). The expansion process is implemented by HAMIS module (HAMIS stands for "Hierarchically Agglomerative Method for Improving NPS"). HAMIS works by calculating classification results on datasets first, then based on them it calculates semantic similarity distance between each pair of clients is found. The clients are semantically similar if the classifier run on their datasets gives similar results in terms of recognizing different categories of customers (Promoters, Passives, Detractors). The distance matrix (between each pair of clients) serves for building a structure called dendrogram, which represents outcome of hierarchical clustering algorithm. Clustering is a machine learning method that groups objects (clients) according to a predefined metric (semantic similarity distance). The dendrogram is a tree-based structure which visually represents which clients are close (similar) to each other. The closer they are in a tree structure, the more similar they are. Based on clusters (formed in a dendro-

© Springer Nature Switzerland AG 2020
K. Tarnowska et al., *Recommender System for Improving Customer Loyalty*,
Studies in Big Data 55, https://doi.org/10.1007/978-3-030-13438-9_5

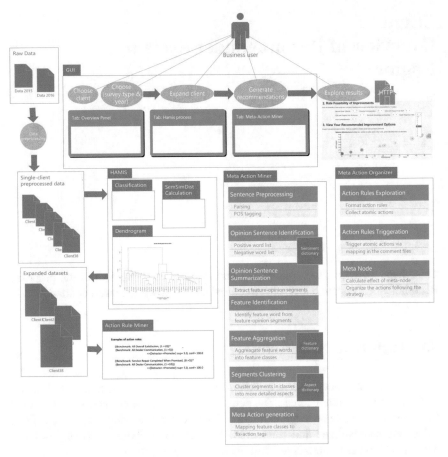

Fig. 5.1 Architecture and main use-cases of the data-driven and user-friendly customer loyalty improvement recommender system

gram structure) expanded datasets are generated (by procedure defined as "climbing the dendrogram" in a bottom up manner). The expanded datasets serve as an input for the action rule mining module. Action rules show which actions should be undertaken to transform customers being labeled unfavorably (such as Detractor or Passive) to a favorable state (in our case-Promoter). The logic behind using expanded datasets for action rule mining is that the worse-performing clients can learn from knowledge gathered by the better-performing clients. The actions are extracted in the form of the changes in the benchmark score (for example, Benchmark: "Service-Repair Completed When Promised, $5 \rightarrow 8$"). These "atomic" actions have to be triggered by some higher-level events, which we call meta-actions.

Meta actions are mined in the Meta-Action Miner tab. The process is implemented in the Meta Action Miner module as a text mining process extracting sentiment from text comments complimentary to surveys. Meta Action Organizer module implements the mechanism of triggering action rules by meta actions. The former process consists of text preprocessing: parsing and POS tagging, extracting sentiment based on opinion words list, aspect-based sentiment extraction (based on pre-defined dictionary of features and their aspects), text summarization (clustering into actual meta-actions). Finally, meta-nodes are created from different combinations of meta-actions. Based on the impact of the meta-actions triggering action rules (and the same scope of changing customers into Promoters) the NPS effect for each meta-node is also calculated. Meta-nodes with the largest effect are considered as recommendations of a system to the user to improve NPS.

The final results of mining are displayed in a web interface (or optionally in the overview panel) in the form of recommendable items. Each item consists of a set of options introduced as changes to improve NPS. Each item is characterized by an attractiveness, derived from the NPS Impact (calculated by the system based on mining results) and from the feasibility of particular changes as stated by the user based on a current clients situation and capabilities of introducing the changes. The items are displayed as bubbles located and colored in a chart according to their attractiveness. Each item can be further analyzed by reviewing actions it suggests to introduce and all the raw comments (divided into positive and negative) associated with the area (aspect) of the action.

5.2 Data Preparation

Usually, much of the raw data contained in the databases is unpreprocessed, incomplete, and noisy [1]. Examples include:

- Fields that are obsolete,
- Missing values,
- Outliers,
- Data in a form not suitable for the data mining models,
- Values not consistent with policy or common sense.

The raw data available for this research is collected by interviewers conducting telephone surveys with customers. The data is inserted into forms manually by interviewers. Therefore, as each manually entered data it is prone to mistakes, inconsistencies and typing errors. Thus, the data should be preprocessed to make it useful for data mining and consequently yield reliable results. The preprocessing includes data cleaning and data transformation.

5.2.1 Raw Data Import

The data is highly multidimensional. There are different types of surveys—questions asked for each type of survey are different, as well as set of asked questions vary between companies. Also, questions change in surveys year by year or even within

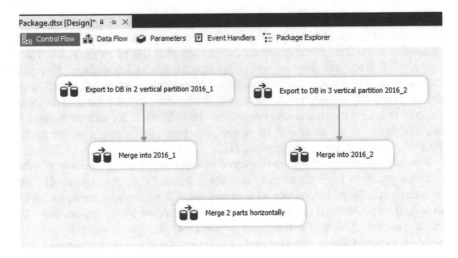

Fig. 5.2 Control flow task definition for the process of importing the data

a year. Consequently, features (benchmarks) are not stable. This results in dataset complexity. The data from the consulting company's database is retrieved into spreadsheet files by means of dynamic reports: the report pulls whatever benchmarks were added as additional columns. If the benchmark has no answers for any of the records in the selection criteria, it is not added to the report.

The customer dataset received from the year 2016 included all the transactions within the specified date range. The raw data was received in two parts, because of the data volume that a single Excel file could not handle.

For further data preprocessing it was necessary to merge both parts and import the data into the SQL database. The first data part (from January to June) had 32,169 transactions, while the second part (June–December 2016)—45,905. The dimensionality of the former-485, the latter-638 columns. The total dataset consists of 78,074 records and 638 columns. The first problem related to importing data from Excel was that SQL Import tool does not handle tables with more than 255 columns. The second problem was related to merging tables from different halves of the year, that is, tables with different number of columns. As a solution, SQL Server Integration Services was used to develop complex and custom ETL (Extraction, Transformation, Loading) processes. The first problem mentioned above was solved by means of dividing data vertically, importing them to the database and merging data again vertically. The second problem was solved by merging data horizontally (see Fig. 5.2 for the Control Flow of the tasks defined in SSIS).

Firstly, for each Excel source file names were defined for accordingly importable chunk of spreadsheet. For the first excel source file (485 columns) the names were defined for:

- Columns from 1 to 255
- Columns from 256 to 500

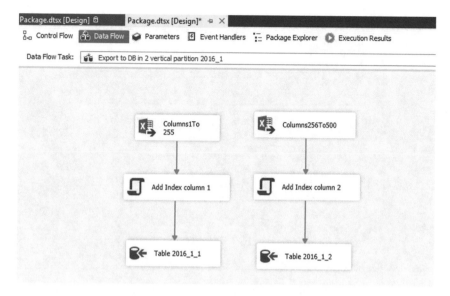

Fig. 5.3 Data flow task definition for the process of exporting data from Excel to database in two vertical partitions (division based on columns) and adding Index column

For the second Excel file (with 638 columns), three names were defined:

- Columns from 1 to 255
- Columns from 256 to 500
- Columns from 500 above

Each name within each Excel source file was defined as a separate source in a SSIS Data Flow task (see Fig. 5.3). Next, the data divided by columns was merged by means of artificially created Index for each row. The additional Index column was created with Script Component. The Merge Join transformation combined all columns into one table for each record based on the Index column (see Fig. 5.4). As a result, two intermediate tables were created for database: 2016_1 (for the first half of 2016) and 2016_2 (for the second half of 2016). In the last Data Flow task both tables were combined into one final table for the year 2016, taking into account different number of columns (Union All component was used for combining rows from both sources)-see Fig. 5.5.

5.2.2 Data Preprocessing

Preprocessing steps typically include:

- Data cleaning
- Handling missing data

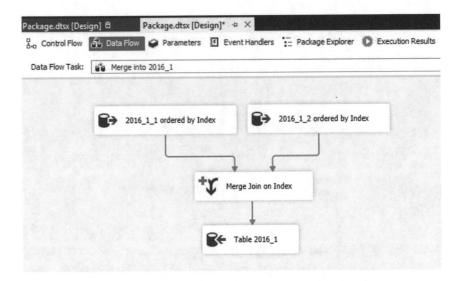

Fig. 5.4 Data flow task definition for the process of merging rows based on Index column

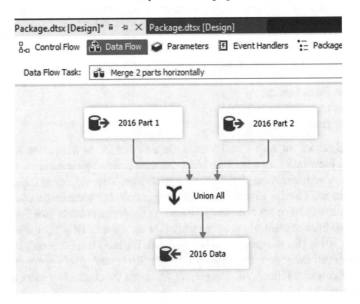

Fig. 5.5 Data flow task definition for the process of combining two parts of 2016 into one final table (merging data horizontally with UNION ALL)

- Transforming into numerical variables

Initially, all preprocessing steps were performed manually by means of scripts (see Fig. 5.6). To automatize all data preprocessing, a package in SQL Server Integration Services was used (Fig. 5.7). Each of the following preprocessing steps were develop

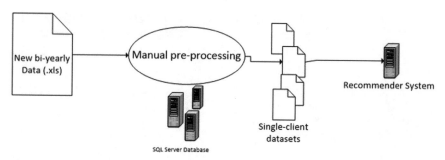

Fig. 5.6 Initial manual preprocessing of the data before loading it into recommender system

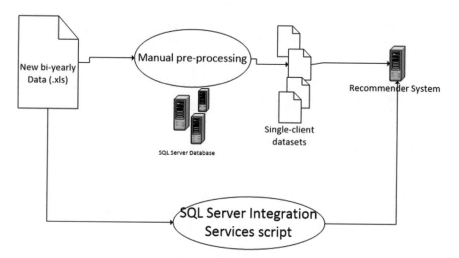

Fig. 5.7 Diagram showing automation of the data preprocessing that can be also integrated into recommender system

by means of Execute T-SQL Statement task, which basically allows to execute any T-SQL script. With the package it is possible to execute all preprocessing steps sequentially (Fig. 5.8). The package can also be executed from Java environment, which makes another step into automatizing the whole recommender system.

1. **Preprocessing Date Column** - transforming date columns (*DateInterviewed, InvoiceDate, WorkOrderCloseDate*) to numbers. Features in a date format (mm-dd-yy or yy-mm-dd) are difficult for machine learning algorithms to handle. Therefore, the dates were transformed to numbers by counting the days from one specified day (we use 12-31-2014 as the interview dates are for year 2016, but some invoice date happened to be in 2015). The new numerical values for the new columns denote which consecutive day starting from first day of 2015 it was.

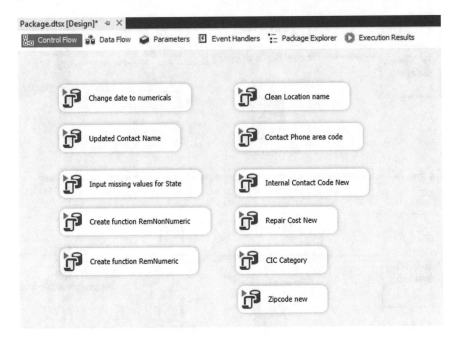

Fig. 5.8 Preprocessing steps implemented in MS SQL Server Integration Services

2. **Updated Contact Name** is a column that contains the new updated contact person name. Based on that column, a new column *UpdatedContactNameNew* was added. It informs whether a contact has been updated, with value "Yes" denoting that contact name has been updated (when *Updated Contact Name* contains a value), and "No" denoting that the contact name has not been updated (when *Updated Contact Name* is empty). Again, this binary column is much more convenient for algorithms to handle than practically unique values of columns Contact Name / Updated Contact Name.

3. **Imputing Missing Values for the Column 'State'** based on data in *Zip* and *Contact Phone* (area code) columns. The column *State* originally was missing values for 35,457 rows (about 45%). After applying the script on completing the data based on the Zip column (which was missing for only 4,791 rows), the data incompleteness for 'State' shrank to only 94 records (0.12%). The new column with the inputted values was called *State Update*.

4. **Cleaning the Values in Location Column** for example, some values have erroneously numbers in their names.

5. **Cleaning the Values in *Internal Contact Code* Column**, adding "ICC" at the beginning.

6. **Extracting Area Code from Contact Phone Column**. First, cleaning is done on *Contact Phone* column, removing all non-numeric data. As a standard, phone should

be represented as ten digits, but there are different variants of the notation in the databases including parentheses for the first three digits, dashes between sequences of digits, etc. However, given the uniqueness of each phone number, standardization of format is not sufficient. We need to decrease selectivity of the column, and therefore it was decided to keep only three first digits which retain information about customers locations by means of area codes. Therefore, based on area code, customers from the same area can be identified. The new column *Contact Phone area code* has categorical values, and adds letters "CP" at the beginning if the area code was derived from Contact Phone or "UP" if it was derived from *Updated Phone* column.

7. **Repair Cost** completing the column with value "0" if it is empty.

8. **Cleaning the Zip Column** by keeping only first 5 digit (format for *Zip* column varies in original data).

9. **Cleaning the CIC Code and PWC Code** columns from numerical values defining *CIC category* column as first two letters from the codes.

10. **Adding Column on County**, based on *Zip code* and *Contact Phone Area Code*. Cleaned and transformed data can be further used for data mining tasks, such as classification and action rule mining.

In summary, the benefits of the presented data import and preprocessing solution with the use of SQL Server Integration Services and tested on 2016 data are as follows:

- A package implemented once, can be executed many times,
- ETL processes can be customized and modified according to the needs,
- SSIS offers validity check,
- A package can be executed from the Java environment and therefore can be integrated into Recommender System.

5.3 Semantic Similarity

The dataset was divided into single-client subsets (38 in total). Additional attributes were developed, including spacial and temporal attributes.

In the first place, classification experiments were conducted for each single dataset in order to determine the predictive capability of a standard classifier model and the same ability to discern and recognize different types of customers (Promoters, Passives and Detractors). It was discovered that the classifier's accuracy/coverage was high for the category "Promoters", but low for the two other categories "Passives" and "Detractors".

RSES (Rough Set Exploration System [2]) was used to conduct initial experiments. The results of the classification experiments—accuracy, coverage and confusion matrix, for Service data for each client were implemented into a visualization system (in this chapter).

Following the classification experiments, the notion of semantic similarity was defined [3]. Assuming that RC[1] and RC[2] are the sets of classification rules extracted from the single-client datasets (of clients *C1* and *C2*), and also:

$RC[1] = RC[1, Promoter] \cup RC[1, Passive] \cup RC[1, Detractor]$,   where the above three sets are collections of classification rules defining correspondingly: "Promoter", "Passive" and "Detractor":

$RC[1, Promoter] = \{r[1, Promoter, i] : i \in I_{Pr}\}$
$RC[1, Passive] = \{r[1, Passive, i] : i \in I_{Ps}\}$
$RC[1, Detractor] = \{r[1, Detractor, i] : i \in I_{Dr}\}$

In a similar way we define: $RC[2] = RC[2, Promoter] \cup RC[2, Passive] \cup RC[2, Detractor]$.

$RC[2, Promoter] = \{r[2, Promoter, i] : i \in J_{Pr}\}$
$RC[2, Passive] = \{r[2, Passive, i] : i \in J_{Ps}\}$
$RC[2, Detractor] = \{r[2, Detractor, i] : i \in J_{Dr}\}$
By $C1[1, Promoter, i]$, $C1[1, Passive, i]$, $C1[1, Detractor, i]$ we mean confidences of corresponding rules in a dataset for client $C1$.

We define $C2[1, Promoter, i]$, $C2[1, Passive, i]$, $C2[1, Detractor, i]$ as confidences of rules extracted from $C1$ calculated for $C2$.

Analogously, $C2[2, Promoter, i]$, $C2[2, Passive, i]$, $C2[2, Detractor, i]$ are confidences of rules extracted from $C2$, and $C1[2, Promoter, i]$, $C1[2, Passive, i]$, $C1[2, Detractor, i]$ are confidences of rules extracted from client $C2$ calculated for client $C1$.

Based on the above, the concept of semantic similarity between clients $C1$, $C2$, denoted by $SemSim(C1, C2)$ was defined as follows:

$$SemSim(C1, C2) =$$
$$\frac{\sum \{C1[1, Promoter, k] - C2[1, Promoter, k] | k \in I_{Pr}\}}{card(I_{Pr})}$$
$$+ \frac{\sum \{C1[1, Passive, k] - C2[1, Passive, k] | k \in I_{Ps}\}}{card(I_{Ps})}$$
$$+ \frac{\sum \{C1[1, Detractor, k] - C2[1, Detractor, k] | k \in I_{Dr}\}}{card(I_{Dr})}$$
$$+ \frac{\sum \{C2[2, Promoter, k] - C1[2, Promoter, k] | k \in I_{Pr}\}}{card(J_{Pr})}$$
$$+ \frac{\sum \{C2[2, Passive, k] - C1[2, Passive, k] | k \in I_{Ps}\}}{card(J_{Ps})}$$
$$+ \frac{\sum \{C2[2, Detractor, k] - C1[2, Detractor, k] | k \in J_{Dr}\}}{card(J_{Dr})}$$

The metric is used to find clients similar to a current client in semantic terms. It calculates the distance between each pair of clients. The smaller the distance is, the more similar the clients are. The resulting distance matrix serves as an input to the hierarchical clustering algorithm. The output of the algorithm is a structure, called dendrogram.

5.4 Hierarchical Agglomerative Method for Improving NPS

Hierarchical Agglomerative Method for Improving NPS (HAMIS) was proposed in Kuang et al. [4] as a strategy for improving NPS of a company based on its local knowledge and knowledge collected from other semantically similar companies operating in the same type of industry. The strategy is based on the definition of semantic similarity introduced in the previous section. HAMIS is a dendrogram built by using agglomerative clustering strategy and semantic distance between clients.

The dendrogram was visualized in the web-based system by means of a node-link diagram that places leaf nodes of the tree at the same depth (see Fig. 5.9).

The clients (leaf nodes) are aligned on the right edge, with the clusters (internal nodes)—to the left. The data shows the hierarchy of client clusters, with the root node being "All" clients. The visualization facilitates comparing the clients by means of similarity. The nodes that are semantically closest to the chosen client are the leaf nodes on the sibling side. The diagram is interactive: after clicking on the client node, all the semantically similar clients are highlighted with numbers in parentheses denoting sequence of the most similar clients (with 1 - denoting the first most similar client, 2 - the second similar, etc.), and the color strength corresponding to the similarity.

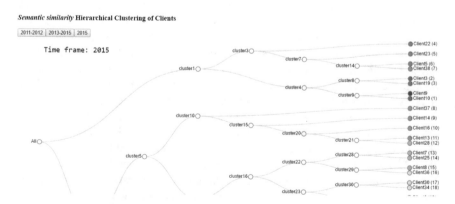

Fig. 5.9 Javascript-based visualization of the dendrogram showing semantic similarity between clients in 2015: chosen Client9 with highlighted semantically similar clients ordered by numbers

The dendrogram was used to construct new "merged" datasets for further data mining (in particular, action rule mining, described in the next section). The merged datasets replace a current client's dataset expanded by adding datasets of better performing clients who are semantically similar to it. So, besides semantic similarity, NPS efficiency rating is another primary measure considered when "merging" two semantically similar clients [4]. As a result of this strategy, the NPS rating of the newly merged dataset will be higher than, or at least equal to, the dataset before its extension. This way, we can offer recommendations to the company with a lower NPS based on the data collected by companies with a higher NPS assuming that these two are semantically similar (that is, their customers understand the concepts of promoter, passive and detractor in a similar way). The second factor considered in the merging operation, besides the NPS, is the quality and consistency of the newly merged data. It is checked by means of F-score calculated for a classifier extracted from the newly merged dataset. The F-score was chosen for keeping track of datasets quality as it combines two other important metrics: accuracy and coverage of the classifier. In summary, three conditions have to be met for the two datasets to be merged:

- merged clients have to be semantically similar within defined threshold;
- NPS of the newly merged dataset must be equal or higher than the original dataset's score;
- F-score of the newly merged dataset must be equal or higher then the currently considered dataset's score.

If these three conditions are met, the datasets are being merged, and correspondingly the current NPS and F-score are updated as well. Then, the merging operation check with the next candidate datasets is continued, until the merging conditions fail or the root of hierarchical dendrogram is reached. By using dendrogram terminology, the current node is being replaced by the newly updated resulting node by "climbing up" the dendrogram. The HAMIS keeps expanding a current client by unionizing it with all the clients satisfying the conditions. The candidates are checked in a top down order based on their depth in the dendrogram: the smaller the depth of a candidate is, the earlier the candidate will be checked. The detailed algorithm for HAMIS and experiments on example runs with it are described in [3, 4]. An example of expanding datasets of 36 clients based on Service 2016 data is shown in Fig. 5.10. Half of the clients were extended by applying the HAMIS procedure, and a client was extended on average by about 3 other datasets. It can be observed that generally clients with lower NPS were extended by a larger number of datasets. It shows that their NPS can be improved more by using additional knowledge from semantically similar, better performing clients. For example, in Fig. 5.10, Client20 with the worst NPS (of 63%) was extended by 10 other datasets and Client33 with the second worst NPS (69%) was extended by 11 other datasets. Expanding the original, single-client datasets was followed by action rule mining–the action rules mined from the extended datasets are expected to be better in quality and quantity. Recommender system based on action rules extracted from the extended datasets can give more promising suggestions for improving clients' NPS score. The more extended the datasets, the better recommendations for improving NPS can be given by the system.

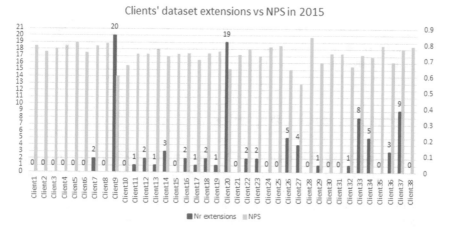

Fig. 5.10 Results of running HAMIS procedure on 38 datasets representing clients for Service survey data from 2016: the number of clients by which a client was extended and its original NPS

5.5 Action Rules

The whole system is built from the knowledge extracted from the preprocessed dataset in the form of action rules. The knowledge is in actionable format and collected not only from the customers using certain business, but also from customers using semantically similar businesses having a higher NPS score.

The concept of action rules was presented in Chap. 3. This kind of actions needs to be extracted, so that an effective recommender system that provides actionable suggestions for improving a client's performance can be built.

The first step to extract action rules from the dataset by the recommender system is to complete the initialization of the mining program by setting up all the variables. This process consists of selecting stable attributes, flexible attributes and the decision attribute. We also need to set up the favorable state and the unfavorable state for the decision attribute, as well as minimum support of the rule and its minimum confidence. *PromoterScore* is set as the decision attribute, with *Promoter* value to be the target state (most favorable one) and *Detractor* the most undesirable state. For the stable attributes, all features related to the general information about clients and customers are considered; the final choice of stable attributes includes:

- *ClientName* - since rules should be client-oriented,
- *Division* - specific department,
- *SurveyType* - type of service: field trips, in-shop, parts, etc.
- *ChannelType*.

Initially, as the flexible attributes, all features denoting numerical benchmark questions were chosen, as it is believed that representing them areas of service/parts can be changed by undertaking certain actions. This set of benchmarks has been reduced to smaller set of benchmarks, which were considered critical. Then, they were used for mining action rules. The choice of critical benchmarks was preceded by an analysis of decision reducts, which are visualized in a user-friendly interface built for the recommender system (described in the next chapter).

5.6   Meta Actions and Triggering Mechanism

The built recommender system is driven by action rules and meta-actions to provide proper suggestions to improve the revenue of companies. Action rules, described in the previous section, show minimum changes needed for a client to be made in order to improve its ratings so it can move to the Promoter's group. Action rules are extracted from the client's dataset expanded by HAMIS procedure, explained in the previous sections.

Meta-actions are the triggers used for activating action rules [5] and making them effective. The concept of *meta-action* was initially proposed in Wang et al. [6] and later defined in Ras et al. [7]. Meta-actions are understood as higher-level actions. While an action rule is understood as a set of atomic actions that need to be made for achieving the expected result, meta-actions are the actions that need to be executed in order to trigger corresponding atomic actions.

For example, the temperature of a patient cannot be lowered if he does not take a drug used for this purpose—taking the drug would be an example of a higher-level action which should trigger such a change. The relations between meta-actions and changes of the attribute values they trigger can be modeled using either an influence matrix or ontology.

An example of an influence matrix is shown in Table 5.1 [3]. It describes the relations between the meta-actions and atomic actions associated with them. Attribute a denotes stable attribute, b - flexible attribute, and d - decision attribute. $\{M_1, M_2, M_3, M_4, M_5, M_6\}$ is a set of meta-actions which hypothetically trigger action rules. Each row denotes atomic actions that can be invoked by the set of meta-actions listed in the first column. For example, in the first row, atomic actions $(b_1 \rightarrow b_2)$ and $(d_1 \rightarrow d_2)$ can be activated by executing meta-actions M_1, M_2 and M_3 together. In the considered domain, it is assumed that one atomic action can be invoked by more than one meta-action. A set of meta-actions (can be only one) triggers an action rule that consists of atomic actions covered by these meta-actions. Also, some action rules can be invoked by more than one set of meta-actions.

If the action rule $r = (a, a_2) \wedge (b, b_1 \rightarrow b_2) \implies (d, d_1 \rightarrow d_2)$ is to be triggered, we consider the rule r to be the composition of two association rules r_1 and r_2, where $r_1 = (a, a_2) \wedge (b, b_1) \implies (d, d_1)$ and $r_2 = (a, a_2) \wedge (b, b_2) \implies (d, d_2)$. The rule r can be triggered by the combination of meta-actions listed in the first and second row in Table 5.1, as meta-actions $\{M_1, M_2, M_3, M_4\}$ cover all required atomic actions: (a, a_2), $(b, b_1 \rightarrow b_2)$, and $(d, d_1 \rightarrow d_2)$ in r. Also, one set of meta-actions can potentially trigger multiple action rules. For example, the mentioned meta-action set $\{M_1, M_2, M_3, M_4\}$ triggers not only rule r, but also another rule, such

Table 5.1 Sample meta-actions influence matrix

	a	b	d
$\{M_1, M_2, M_3\}$		$b_1 \rightarrow b_2$	$d_1 \rightarrow d_2$
$\{M_1, M_3, M_4\}$	a_2	$b_2 \rightarrow b_3$	
$\{M_5\}$	a_1	$b_2 \rightarrow b_1$	$d_2 \rightarrow d_1$
$\{M_2, M_4\}$		$b_2 \rightarrow b_3$	$d_1 \rightarrow d_2$
$\{M_1, M_5, M_6\}$		$b_1 \rightarrow b_3$	$d_1 \rightarrow d_2$

as $(a, a_2) \wedge (b, b_2 \rightarrow b_3) \implies (d, d_1 \rightarrow d_2)$, according to the second and fourth row in Table 5.1, if such rule was extracted.

The goal is to select such a set of meta-actions which would trigger a larger number of actions and the same bring greater effect in terms of NPS improvement. The effect is quantified as following [3]: supposing a set of meta-actions $M = \{M_1, M_2, \ldots, M_n : n > 0\}$ triggers a set of action rules $\{r_1, r_2, \ldots, r_m : m > 0\}$ that covers objects in a dataset with no overlap. The coverage (support) of M is defined as the summation of the support of all covered action rules. That is, the total number of objects that are affected by M in a dataset. The confidence of M is calculated by averaging the confidence of all covered action rules:

$$sup(M) = \sum_{i=1}^{m} sup(r_i)$$

$$conf(M) = \frac{\sum_{i=1}^{m} sup(r_i) \cdot conf(r_i)}{\sum_{i=1}^{m} sup(r_i)}$$

The effect of applying M is defined as the product of its support and confidence: $(sup(M) \cdot conf(M))$, which is a base for calculating the increment of NPS rating.

5.7 Text Mining

Triggers aiming at different action rules are extracted from respectively relevant comments left by customers in the domain [5]. Text comments are a complementary part of structured surveys. For example, for a rule described by: $r = [(a, a_2) \wedge (b, b_1 \rightarrow b_2)] \Rightarrow (d, d_1 \rightarrow d_2)]$, where a is a stable attribute, and b is a flexible attribute, the clues for generating meta-actions are in the comments of records matching the description: $[(a, a2) \wedge (b, b1) \wedge (d, d1)] \vee [(a, a2) \wedge (b, b2) \wedge (d, d2)]$.

Mining meta-actions consists of four characteristic steps involving sentiment analysis and text summarization [8], as described in Chap. 3 in greater detail:

1. Identifying opinion sentences and their orientation with localization;
2. Summarizing each opinion sentence using discovered dependency templates;
3. Opinion summarizations based on identified feature words;
4. Generating meta-actions with regard to given suggestions.

The whole process of mining customers comments uses sentiment analysis, text summarization and feature identification based on guided folksonomy (domain specific dictionaries are built). It also generates appropriate suggestions, such as meta-actions, which is important for the purpose of recommender system.

The schema of the presented aspect-based sentiment mining was inspired by a process described in [9]. *Sentiment analysis* is generally defined as analyzing people's opinions, sentiments, evaluations, attitudes, and emotions from written language. *Aspect-based sentiment analysis* is based on the idea that an opinion consists of a sentiment (positive or negative) and a *target* of the opinion, that is, a specific aspect or feature of the object. It offers more detailed and fine-grained analysis than document-level or sentence-level sentiment analysis. The details are presented in Chap. 3.

Consequently, the first step in text mining consists of identifying an opinion sentence, based on the occurrence of an opinion word. A dictionary (list) of positive and negative words (adjectives) were used for that purpose. Context (localization) was also taken into account. For example, a comment *"the charge was too high"*, "high" is recognized according to the adjective lists as neither positive nor negative. However, the comment still presents an insightful opinion about discontent when it comes to pricing. Therefore, "high" was added to the list as a negative in the context of pricing.

In the next step, sentences with opinion words identified are shortened into segments. Feature-opinion pairs are generated based on grammatical dependency relationships between features and opinion words. The foundation of this step is the grammatical relations defined by Stanford Typed Dependencies Manual [10] and generated by Stanford Parser. A dependency relationship describes a grammatical relation between a governor word and a dependent word in a sentence. Given the wide definition of dependency templates (about 50 defined dependencies in [10]), all the necessary relations associated with opinion words can be identified. On top of it, negation and 'but'-clauses are identified.

Having extracted segments, feature words are identified using the supervised pattern mining method (similarly as described in [11]). The Parts-of-Speech tags (POS) help in the process of recognizing the features.

Opinion summarizations are used in many sentiment analysis works (first in [9]) to generate a final review summary about the discovery results on feature and opinions mining and also rank them according to their appearances in the reviews.

In this work, the focus was also on removing the redundancy of extracted segments and clustering segments into different classes. The feature clustering was based on the pre-defined list of seed words or phrases. To cluster a segment into the corresponding class, its feature word or the base form of its feature is checked whether it exists in any list of the seed words.

For the purpose of generating meta-actions, each feature class has been divided into several subclasses. Each subclass is related to the specific aspect of that feature. The aspects have been defined based on the domain knowledge.

The last step is generating meta-actions and providing them to the end business user along with the comments from which they were mined. The recommendations are divided into positive and negative recommendations. Negative opinions show the undesirable behaviors that should be fixed, while the positive segments indicate which areas should be continued.

References

1. D. T. Larose and C. D. Larose. *Data Mining and Predictive Analytics*. Wiley Publishing, 2nd edition, 2015.
2. RSES 2.2 User's Guide. http://logic.mimuw.edu.pl/~rses.
3. J. Kuang. *Hierarchically structured recommender system for improving NPS*. PhD thesis, The University of North Carolina at Charlotte, 2016.
4. J. Kuang, Z. W. Raś, and A. Daniel. *Hierarchical Agglomerative Method for Improving NPS*, pages 54–64. Springer International Publishing, Cham, 2015.
5. J. Kuang and Z. W. Ras. In search for best meta-actions to boost businesses revenue. In *Flexible Query Answering Systems 2015 - Proceedings of the 11th International Conference FQAS 2015, Cracow, Poland, October 26–28, 2015*, pages 431–443, 2015.
6. K. Wang, Y. Jiang, and A. Tuzhilin. Mining actionable patterns by role models. In L. Liu, A. Reuter, K.-Y. Whang, and J. Zhang, editors, *ICDE*, page 16. IEEE Computer Society, 2006.
7. A. A. Tzacheva and Z. W. Ras. Association action rules and action paths triggered by meta-actions. In *2010 IEEE International Conference on Granular Computing, GrC 2010, San Jose, California, USA, 14–16 August 2010*, pages 772–776, 2010.
8. J. Kuang, Z. W. Ras, and A. Daniel. *Personalized Meta-Action Mining for NPS Improvement*, pages 79–87. Springer International Publishing, Cham, 2015.
9. M. Hu and B. Liu. Mining and summarizing customer reviews. In *Proceedings of the Tenth ACM SIGKDD International Conference on Knowledge Discovery and Data Mining*, KDD '04, pages 168–177, New York, NY, USA, 2004. ACM.
10. M.-C. D. Marneffe and C. D. Manning. Stanford typed dependencies manual, 2008.
11. B. Liu. Sentiment analysis and subjectivity. In *Handbook of Natural Language Processing, Second Edition. Taylor and Francis Group, Boca*, 2010.

Chapter 6
Visual Data Analysis

Visual techniques were used to support complex customer data analytics and illustrate concepts and data mining algorithms behind the built Recommender System. The implemented web-based visual system supports a feature analysis of 38 client companies for each year between 2011–2016 in the area of customer service (divided into shop service/field service) and parts. It serves as a visualization for the feature selection method showing the relative importance of each feature in terms of its relevance to the decision attribute—Promoter Status, and is based on an algorithm that finds minimal decision reducts. It allows the user to interactively assess the changes and implications onto predictive characteristics of the knowledge-based model. It is supported by visual additions in form of charts showing accuracy, coverage and confusion matrix of the model built on the corresponding, user-chosen dataset.

6.1 Decision Reducts as Heatmap

For a basic attribute analysis a heatmap was proposed, which bears similarity to a correlation matrix—see Fig. 6.1. However, here attribute relevance to the decision attribute (Promoter Score) is defined here by means of a reduct's strength (a percentage occurrence of a benchmark attribute in reducts).

The analyst can choose the Client Category (analysis supported for 38 clients and All the clients) and Survey Type category (Service/Service:Field/Service:Shop/Parts). The columns correspond to the benchmarks (that is surveys' attributes) found in reducts (the full benchmark name is visible after hovering the mouse over the benchmark code). The rows represent years, in which customer satisfaction assessment was performed (current version supports datasets from years 2011–2015).

The cells represent benchmark strength in a given year—the color linear scale corresponds to a occurrence percentage. The darker cells indicate benchmarks that belong to more reducts than benchmarks represented by the lighter cells—the darker the cell, the stronger the impact of the associated benchmark on promoter score

© Springer Nature Switzerland AG 2020
K. Tarnowska et al., *Recommender System for Improving Customer Loyalty*,
Studies in Big Data 55, https://doi.org/10.1007/978-3-030-13438-9_6

Customer Sentiment Analysis **Net Promoter Score**

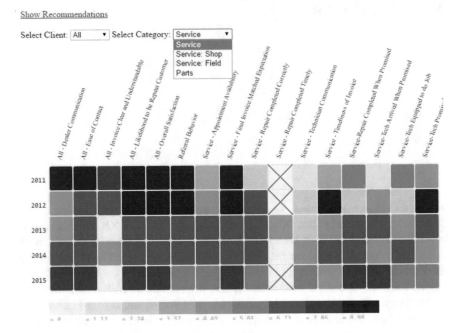

Fig. 6.1 Dynamic visualization of a reduct matrix based on a heatmap design

(color denotes the strength of what 'statement of action rules are saying'). The average benchmark score is visible after hovering over a cell. Additionally, we use red-crossed cells to denote benchmarks that were not asked as questions in a given year for a chosen client so that we do not confuse benchmarks' importance with benchmarks' frequency of asking.

The design allows to track which new benchmarks were found in reducts in relation to the previous year, which disappeared, which gained/lost in strength.

The benchmark matrix should be analyzed together with the NPS row chart to track how the changes in benchmarks' importance affected NPS (decision attribute) changes (see Fig. 6.2). The row chart showing yearly NPS changes per client is complementary to the reducts' matrix. Further, "Stacked area row chart" was added to visualize distribution of different categories of customers (Detractors, Passives, Promoters) per year (Fig. 6.2). It is complementary to the two previous charts, especially to the row NPS chart and helps to understand changes in Net Promoter Score. The distribution is shown based on percentage values, which are visible after hovering over the corresponding area on the chart. We used colors for differentiating categories of customers as analogy to traffic lights: red means Detractors ('bad' or 'angry' customers), yellow for Passives, and green for Promoters.

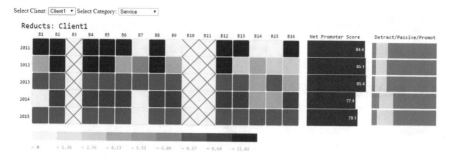

Fig. 6.2   Reduct heatmap with NPS row chart and NPS category distribution chart

6.2   Classification Visualizations: Dual Scale Bar Chart and Confusion Matrix

For the purpose of visualizing classification results dual scale bar chart was used (Fig. 6.3). It allows to additionally track knowledge losses by means of both: "Accuracy" and "Coverage" of the classifier model trained on single-client datasets. It is interactive and updates after the user chooses a "Client" from drop-down menu.

Additionally, confusion matrix was used to visualize classifier's quality for different categories (see Fig. 6.3). From the recommender system point of view it is important that the classifier can recognize customers (covered by the model), because it means their sentiment can be changed. The chart updates after interacting with the dual scale bar classification chart (hovering over the bar for the corresponding year) and shows detailed information on the classifier's accuracy per each year. The rows in this matrix correspond to actual decision classes (all possible values of a decision), while columns represent decision values as returned by classifier in discourse. The values on diagonal represent correctly classified cases. Green colors were used to represented correctly classified cases, while red scale to represent misclassified

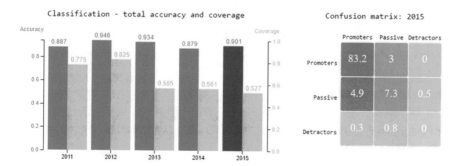

Fig. 6.3   Visualizations for the classifier's results—accuracy, coverage and confusion matrix

cases. The color intensity corresponds to the extent of the confusion or the right predictions.

6.3   Multiple Views

Multiple views design was used to enable a multidimensional problem analysis. However, this was quite challenging to design, as such view requires a sophisticated coordination mechanism and layout. The goal was to balance the benefits of multiple views and the corresponding complexity that arises.

The multiple view was designed to support deep understanding of the dataset and the methodology for recommendations to improve NPS. The basic idea behind it is based on hierarchical structure of the system and a dendrogram was also used as an interface to navigate to more detailed views on a specific client. All other charts present data related to one client or aggregated view ("All"):

- Reducts matrix (Fig. 6.1) is a way to present the most relevant attributes (benchmarks) in NPS recognition process in temporal aspect (year by year),
- NPS and Detractor/Passives/Promoters charts (Fig. 6.2) are complementary to reducts matrix and help track changes in Net Promoter Score in relation to the changes of the benchmark importance,
- Detractor/Passives/Promoter distribution chart helps understand nature behind NPS changes and customer sentiment change trends,
- Classification and confusion matrix charts (Fig. 6.3) help tracking the quality of gathered knowledge of NPS in datasets within different years and understand the reasons for knowledge loss (seen as decrease in classification model accuracy) in terms of benchmark changes (survey structure changes),
- Confusion matrix is a more detailed view of classification accuracy chart and helps understanding the reasons behind the changes in the accuracy yearly along with the customer's distribution chart.

In conclusion, the problem of changing datasets per year and per client and therefore changing the model and the algorithm can be fully understood after considering all the different aspects presented on each chart.

6.4   Evaluation Results

In the following subsection the evaluation of the visual system on the chosen use-case scenarios is described.

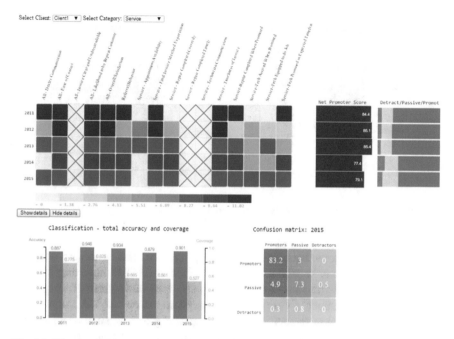

Fig. 6.4 Visual multiple analysis for a single client ("Client1")

6.4.1 Single Client Data (Local) Analysis

A case study of analysis process is presented here on the example of the chosen Client 1 (see Fig. 6.4). In the first row of the reduct matrix only dark colors can be observed, which means that all the colored benchmarks are equally high important. Customers' sentiment is pretty certain and defined on benchmarks. Next year, importance of benchmarks dropped a lot, the values decreased. For some reasons, customers changed their opinion. Average score of these benchmarks says that customers are little less satisfied. New benchmarks are showing up, but they are not that important. Some benchmarks lost importance changing the color from the darkest to lighter. We can observe movement of people from Detractors to Promoters, in connection with benchmark analysis it provides much wider view. Checking confusion matrix allows to go deeper into what customers are thinking. In the third year all benchmark lost in their way. Customers do not have strong preference on which benchmarks are the most important, but NPS is still going up. In the fourth year 2 benchmarks disappeared, in fifth year all of benchmarks are again getting equal strength. One can observe a huge movement into Detractor group in 2014, but at the same time customers got more confidence which benchmarks are the most important. In 2015 all of the benchmarks have the same strength.

Number of Passives increases from 2014 to 2015, and the move to this group is from Detractor group (we could observe a huge movement from Detractors to

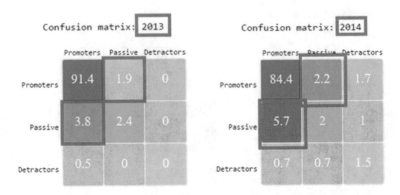

Fig. 6.5 Classification results analysis—confusion matrices

Passives). Net Promoter Score was moving up and later the customers were not certain (about benchmarks preference) and then there followed a huge drop in NPS.
  Looking at the classification model, coverage jumped down from 2013 to 2014 incredibly (that is number of customers which system can classify). Accuracy went down, and by looking at the confusion matrix (see Fig. 6.5) one can conclude that more people who are similar in giving their rates are confused by the system between Promoters and Passives. Some of such customers end up in Promoter category, while the others in Passives.

6.4.2   *Global Customer Sentiment Analysis and Prediction*

By looking at the "Reducts" heatmap, a conclusion can be drawn that the customers' sentiment towards the importance of benchmarks is apparently changing year by year. Some benchmarks are getting more important, some less. By looking at the benchmarks for "All" companies (see Fig. 6.1), "Overall Satisfaction" and "Likelihood to be Repeat Customer" seem to be the winners. Clearly, it will differ when the personalization goes to the companies' level. It might be interesting to personalize further—going to different seasons (Fall, Spring, etc.) and seeing if the customers' sentiment towards the importance of benchmarks is changing. The recommendations for improving NPS are based on the customers' comments using the services of semantically similar companies (clients) which are doing better than the company we want to help. So, it is quite important to check which benchmarks they favor. It is like looking ahead how our customers most probably will change their sentiment if the company serving these customers will follow the recommendation of the system we build. In other words, it is possible to somehow control the future sentiment of the customers and choose the one which is the most promising for keeping NPS improving instead of improving it and next deteriorating.

From the proposed and developed charts, one can see how customers' sentiment towards the importance of certain benchmarks is changing year by year. Also, the business user can do the analysis trying to understand the reasons behind it with a goal to prevent them in the future if they trigger the decline in NPS.

The advantage of using recommender system the way it is proposed - based on hierarchical structure (dendrogram), assuming that companies use its recommendations, will be the ability to predict how customers' sentiment towards the importance of certain benchmarks will probably change with every proposed recommendation. So, one can not only pick up a set of benchmark triggers based on their expected impact on NPS but also on the basis of how these changes most probably will impact the changes in customers' sentiment towards the importance of benchmarks. This way a tool can be built which will give quite powerful guidance to companies. It will show which triggers are the best, when taking into account not only the current company's performance but also on the basis of the predicted company's performance in the future assuming they follow the system recommendations.

6.5 User-Friendly Interface for the Recommender System

For review summarization purposes, often a variety of visualization methods are deployed in the literature. Some of them were presented in Chap. 3. For the recommender system, as described in the previous chapter, an interactive user-friendly web-based interface was built to summarize the results. The interaction was divided into three basic steps:

1. Selecting the entity (client) the business user would like to analyze (see Fig. 6.6);
2. Rating the feasibility of improvements (drop-down lists in Fig. 6.7);
3. Exploring the recommended improvement options (bubble chart in Fig. 6.7) and comments from raw data related to the chosen option (data table in Fig. 6.8).

The map in Fig. 6.6 serves as an interface for further analysis of the chosen client (amongst 38 in total). The current version of the interface allows for choosing recommendations based on the datasets from the years 2016 or 2015 and surveys on Service or Parts. The clients are represented as points (dots) placed in their headquarters' locations. The size of the dot informs about how many other clients were added to the original client's dataset to mine for actionable knowledge (see section on semantic similarity and HAMIS procedure). The connecting lines show the semantic neighbors. After clicking the client's dot, it changes color from blue to red and the corresponding semantic neighbors are highlighted in a red scale as well. The color scale corresponds to the strength of semantic similarity. Additionally, the number in parentheses denotes the sequence of semantic similarity to the current client. The client labels (text next to the dots) have been hidden on the grounds of data confidentiality.

The next step of interaction with the business user is exploring the recommendation options. The displayed options correspond to the extracted meta-actions (see the

Fig. 6.6 Javascript-based visualization for depicting clients' locations and their semantic neighbors. Also, serves as an interface for further analysis of a chosen entity

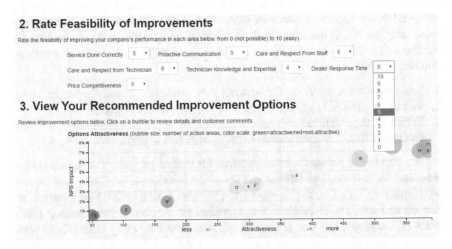

Fig. 6.7 Javascript-based interactive visualization for exploring recommendations options and their attractiveness based on chosen feasibility

previous section) mined from text comments and summarized into aspect categories. The user (business consultant) can assign a feasibility score to each option based on dialogue with the currently analyzed client. For some clients some options might be more feasible that the others. For example, it might be quite difficult to change

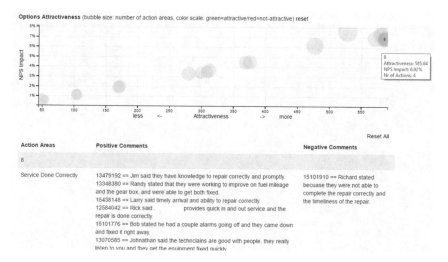

Fig. 6.8 Javascript-based dynamic data table for exploring raw data comments associated with the analyzed recommendation option

pricing, while it might relatively easier to change technician knowledge (for example, by introducing appropriate training).

The option's attractiveness depends on both factors: NPS improvement (calculated as described in the previous section based on the action rule and meta action mining) and feasibility chosen by the user. Each bubble (identified by an ordering number) corresponds to a different set of improvement options and they are ordered on the X-axis and the Y-axis according to their attractiveness (see Fig. 6.7). The most attractive options lie in the top right corner of the chart. The attractiveness is also denoted by the color scale—from red scale(unattractive) to green scale (attractive).

The user can choose the option and analyze it further (see Fig. 6.8): the highlighted bubble shows details on:

- number of actions included in the option;
- the quantified attractiveness (calculated as combination of feasibilities and NPS impact of actions in the option);
- the combined overall NPS impact.

Furthermore, the data table shows raw text comments from customers associated with the particular areas (aspects), divided into negative and positive columns (see Fig. 6.8). Each comment can be investigated further by analyzing the whole survey and context in which it was expressed, as each comment is identified with Work Order ID.

Chapter 7
Improving Performance of Knowledge Miner

7.1 Introduction

One of the most important aspects in the built Miner is to ensure its scalability and high performance. Ideally, it should work in real time, that is, it should update its knowledge base whenever new data is available. On the other hand, the model needs to work on all historical data, therefore each update of the system requires building a model on a complete yearly data. It is known that data mining algorithms run on large amounts of data that take time, which not always can be acceptable for business settings.

The main motivation for taking up the topic of improving miner performance was to ensure the built system's scalability and make the system's knowledge update acceptable for practical business environment.

Within this subtopic, different methods for improving action rule mining were proposed and a number of experiments proving its effectiveness were performed. Also, a final conclusion on which method should be applied in practical settings was delivered.

7.2 Problem Statement

Action rule mining is a major step in extracting knowledge for a data-driven recommender system (as described in Chap. 4).

The system in its current version supports to provide recommendations to 38 companies, but should also be scalable for adding new companies. Furthermore, recommendations can be based on the data from the chosen survey type (for example, for service, products, rentals, etc.) and for the chosen year (the system should update yearly). This broad range of choices creates many cases for which recommendations should be extracted. Each case requires separate knowledge extraction procedure, which taking into account many combinations (company/survey

© Springer Nature Switzerland AG 2020
K. Tarnowska et al., *Recommender System for Improving Customer Loyalty*,
Studies in Big Data 55, https://doi.org/10.1007/978-3-030-13438-9_7

type/year) and datasets' sizes, becomes quite time-consuming. Also the system should be extended by discovering rules of types "Passive to Promoter" and "Detractor to Passive" so that to improve accuracy of recommendations and increase the coverage of customers and addressing their concerns. We identified a need to optimize (speed up) action rule mining as it became a major time-intensive task in the process of updating the system with the new knowledge. It also became a potential obstacle in making the system fully scalable (to an even greater number of companies or to extending the RS to companies branches shops level).

7.3 Assumptions

To summarize, the need for scalability results from the following assumptions for the system:

- The system supports 38 companies currently, but should scale up to more in the future.
- The knowledge base of the system should be updated bi-yearly.
- The system allows to choose recommendations based on different survey sources: surveys for service, parts, rentals, etc.
- The system should allow to choose recommendations based on different time-frames: yearly, 18-months timeframe, etc.

For each combination of the cases mentioned above the mining process has to be done separately.

The action rules extracted for the recommender system have the structure as described in Chap. 3. Action rules applied to the Net Promoter Score show which areas of customer service/products should be improved in order to change a customer labelled as "Detractor" to "Promoter" or at least to "Passive". Also, the goal is to "convert" passive customers to active promoters. The sample action rules are presented in the listing below:

Listing 1 Sample action rules.
(Benchmark: All − Overall Satisfaction, 1−>10) AND (Benchmark:
All − Dealer Communication, (1−>5) => (Detractor−>Promoter) sup=5.0,
conf=100.0

(Benchmark: Service − Repair Completed When Promised, 8−>3) AND
(Benchmark: All − Dealer Communication, (1−>10))
=> (Detractor−>Promoter) sup=5.0, conf=100.0

Each action rule is characterized by support, that is how many customers are matched with the action rule (with the action rule's "before" state), and confidence that says what the probability of "converting" a customer is.

Also, it was observed, that in general, a large number of action rules is extracted. Many action rules are redundant since that target the same customers. On the other hand, there are detractor and passive customers that are not covered by any of the rules.

This means many customers can be changed by different ways, that is many customers are matched by different action rules. An important measure that needs to be kept track of besides the total number of rules, is how many distinct customers are covered by the rules. This directly corresponds to the NPS measure which is calculated as the percentage of Detractors subtracted from the percentage of Promoters.

There are business constraints for the update of the system. One of the non-functional requirements for the system, as defined by the business users, is that assuming the bi-yearly update of the system's base knowledge, the update cannot take longer than 24 h. Considering the number of companies (currently 38), running the mining of action rules and meta-actions should take approximately up to 30 min per company on average using a personal PC machine.

7.4 Strategy and Overall Approach

There are several methods proposed to improve the time efficiency of the action rule mining for the system.

Firstly, the dataset has to be prepared accordingly before mining. For example, the dataset extracted from the database has values "NULL" as imputed for missing values. When not removing these, they are considered by the action rule mining program as a separate value.

Secondly, a limit on the number of extensions for each client was proposed. As already mentioned in the previous chapters, the approach is based on extending a single company's dataset by its nearest semantic neighbors, who perform better in terms of NPS. This intuitively means, that a company can "learn" from a better performing company, which is similar to it, by means of what customers are labeled as either Promoter, Passive or Detractor are saying about customer experience. For example, preprocessing service data for recommendations from the year 2015, resulted in extreme cases (for the clients with the worst NPS) of 19 (Client20) and 20 extensions (Client9)—see Fig. 5.10 in Chap. 5. The original datasets contained 479–2527 rows correspondingly, and after extension they had 16,848–22,328 rows. Action rule mining time is increasing exponentially to the dataset's size.

For 2016 data, the process of extending datasets according to HAMIS strategy resulted in potential improvement of NPS from 76% to 78% on average for 38 companies, but also increased the average size of the datasets from 873 rows to 2159. In case of data for Parts surveys these numbers were, correspondingly, 81–84% of potential NPS improvement, and the side effect was the increase of datasets' sizes from 559 to 2064 on average. Extending such datasets itself takes a long time or even results in "out of memory" error. Such vastly extended datasets used for action rule mining, prolong the process of knowledge extraction used for recommendations by weeks or even months, which is unacceptable from the business point of view.

Limiting extensions to the three nearest neighbors solves the situation for the "extreme" cases of extended dataset, however it does not improve the overall time of action rule mining, which in total was still unacceptable from the business point of view.

The third method targets at decreasing the number of distinct values in the benchmark columns of the datasets. It can be achieved by means of a preprocessing technique known as *binning numerical variables*—that is dividing the current values into ranges and naming the ranges as categorical values, for example: low, medium, high.

The proposed binning method for the benchmark scores, whose scale is 0-10 in most cases, is as follows:

- 0–4: low,
- 5–6: medium,
- 7–8: high,
- 9–10: very high.

The method was implemented by means of the data preprocessing script. The proposed method of creating bins was mostly determined by business understanding of which values mean "low" or "high". The other methods used for binning, in general, are: based on equal width, based on equal frequency, binning by clustering or binning based on predictive value. Also, some benchmarks have scale other than 0–10, for example: 1–2, 0–1 (yes/no benchmarks) or 0–6. Such benchmark attributes were not binned, as they already have low cardinality.

The fourth method was to put constraints on the pattern of mined action rules. The current miner checks for all the combinations of possible values in the atomic action sets. As the listing of a sample action rule above shows, these can also contain actions that suggest changing the value of a benchmark score from higher to lower. However, such changes do not make sense from the business and common sense point of view. They are useless, as the goal is to find actions for clients to undertake the improvement of their benchmark ratings. Therefore, the goal was to restrict mined patterns so that to extract only action rules that suggest changing benchmark scores to higher values (for example, from 8 to 9). Rules containing actions suggesting benchmark's score change from higher to lower (for example, from 8 to 3) or no change in value (for example, from 10 to 10) would be discarded. In further analysis we call such rules "Rule Type 0". The example of such rules is presented in the listing below.

Listing 2 Sample action rule of "Type 0".
((Benchmark: All − Likelihood to be Repeat Customer,
1−>9)) AND (Benchmark: All − Ease of Contact, (5−>6)))
=> (Detractor−>Promoter) sup=2, conf=100.0

The fifth method goes further in changing the pattern of mined rules. The assumption here is that the value on the right hand side of the flexible attribute does not matter. In other words, the actions should suggest changes to any higher values than the left side value. 'Higher' here means 'any value higher than the left-hand value'. As a result, atomic action sets would contain changes, such as: "1→ higher" (means 1→

2 or 3 or ... or 10), "8→ higher" (means 8→ 9 or 10), etc. A sample extracted rule is presented in the listing below.

Listing 3 Sample action rule of " Type 1".
((Benchmark: Service − Tech Equipped to **do** Job,
2−>higher)) AND (Benchmark: Service − Repair Completed When Promised,
(1−>higher))) => (Detractor−>Promoter) sup=2, conf=85.0

These rules are called later "Rule Type 1" to differentiate from the previous pattern defined ("Rule Type 0"). The disadvantage of this method is that the recommender system has to be adapted to work on such rules (read and process into recommendations).

The fourth and fifth methods aim at reducing the number of atomic action sets and therefore improve running time of the mining process. The potential disadvantage here is the loss of certain knowledge and the extracted rules can be less accurate.

The sixth method is based on parallelizing the action rule mining algorithm by implementing it in a distributed manner. This method requires revising and proposing a new distributed algorithm for the action rule mining. Such can be run in a clustered computing environment and take advantage of running in a parallel manner. The method used here used Spark environment and academic cluster.

7.5 Evaluation

The methods proposed in the previous section were implemented and tested by means of various metrics.

The metrics measured within experiments included:

- Running time (measure in milliseconds, displayed in seconds, minutes or hours),
- The number of rules extracted (including duplicates),
- Coverage—understood as the number of distinct customers covered by the rules. This measure is of more interest than the number of rules overall, as it directly affects NPS and the mechanism of rules' triggering (described in Chap. 4). The goal is to maximize the number of different customers whose Promoter Status is to be changed.

Also, different cases were tested with the recommender system—the resulting action rule files from each case were read and processed in the system. Corresponding triggering mechanism was checked along with the final generated recommendations (called meta-nodes) and maximal NPS Impact one could get from them.

7.5.1 Experimental Setup

The base performance was defined as a state before any modifications (before applying any of the methods described above).

The preliminary testing was performed on the 2015 data to get a general outlook on the feasible magnitude of speedup—since 2015 datasets were already mined for action rules and the times were known already.

The main testing was performed on the companies' datasets from 2016. The datasets chosen for performing tests on were divided into different groups taking into account its size:

- Small datasets (about 200–300 rows),
- Medium datasets (about 1,000–1,200 rows),
- Large datasets (about 5,000–7,000 rows),
- Very large—the datasets that were not cut to the three nearest semantic neighbors (about 10,000 rows).

Another test case for each size of the dataset included:

- Non-binned numerical values for benchmark attributes,
- Binned values of benchmark attributes: very low, low, medium, high.

Furthermore, tests were performed on extracting two types of rules, considering the expected change of NPS status on the right hand side of the action rule:

- Detractor$\rightarrow$ Promoter,
- Passive$\rightarrow$ Promoter.

The following test cases, considering the format of extracted action rules, were defined:

- Base rule type: allowing NULLs as values and allowing flexible attributes (benchmarks) to go down or stay the same in an action rule, for example:

Listing 4 Example of a base rule type.
  (Benchmark: Parts − Knowledge of Personnel, 10−>6) AND
  (Benchmark: All − Dealer Communication, (6−>NULL))
  => (Detractor−>Promoter) sup=2, conf=87.5

- Rule "Type 0"—allowing flexible attributes to only go up—this was implemented in the Miner module by restricting atomic action sets to a defined pattern,
- Rule "Type 1"—flexible attributes can go up to any higher value.

Also, the tests were performed taking into account implementation type and hardware/infrastructure setup:

- Non-parallel implementation (Java Miner).
- Distributed implementation of the action rule mining algorithm (implemented in the Spark environment).

7.5.2   Results

The results from the experiments on the test cases and metrics as defined in the section above are summarized in the tables and described in the following subsections.

Table 7.1 A summary of experiments on a small dataset (Client 24 - 306 surveys, rules Detractor to Promoter)

Test case	Running time (s)	Coverage (%)
Base	10	71.4
Type 0	5	71.4
Type 1	2.3	71.4
Binned—base	17	71.4
Binned type 0	4.3	71.4
Binned type 1	3	71.4

7.5.2.1 Small Datasets

The smallest dataset for Service category in 2016 (for Client 5) consisted of 213 surveys and was not large enough to generate any rules of the type "Detractor→ Promoter".

The second smallest dataset (Client 24) involved 306 surveys and the results for different test cases are shown in Table 7.1. For a small dataset, it could be observed that switching the method from the basic rule type to rule "Type 0" decreased running time twice as much. When further changing the rule type to "Type 1", the running time further decreased two times (that is, it was one fourth of the original base time).

Binning numerical variables (benchmarks) seemed to increase running time in comparison to a non-binned small dataset—from 10 s to 17 s. The binned dataset with rule "Type 0" was 4.3 s and "Type 1"—3 s, which were a little bit worse than the corresponding non-binned datasets.

The coverage of customers was the same for all the tested cases –71.4%, although the base case generated much more rules (1342 rules, "Type 0"—329, "Type 1"—141). Apparently many of them were redundant (they did not target more detractors) and this only added to the running time.

In summary, for a small dataset, changing the rule type brought a significant speedup, however binning the datasets did not, contrary to the expectations. Both modifications did not change the coverage of the dataset—for all the cases 5 detractors (out of 7 in total) were targeted by the extracted rules. Still, the small datasets are usually not problematic in terms of running time (it completes within seconds)—however they give first estimates by how much the running time can be decreased.

The small datasets were also tested on "Passive to Promoter" action rules. For this case, the smallest dataset (213 rows) generated results (see Table 7.2). There were 31 passive customers in total for this dataset, and in base case 25 of them were covered by the extracted rules. Changing rule to "Type 0" and "Type 1" decreased number of customers covered down to 23. Binning the datasets also worsened the coverage—24 in base case, 18 for rule types "0" and "1".

It could be observed, that changing rule to "Type 0" decreased running time by about four times, and "Type 1" even a little more. Similar effect was observed for the cases of binned datasets. This time, also binning in comparison to non-binned datasets brought speedup in time (about twice shorter time).

Table 7.2 A summary of experiments on a small dataset (Client 5 - 213 surveys, rules Passive to Promoter)

Test case	Running time (s)	Coverage (%)
Base	77	80.6
Type 0	19.6	74.2
Type 1	15.7	74.2
Binned—base	47.6	77.4
Binned type 0	8.8	58.1
Binned type 1	10.5	58.1

Table 7.3 A summary of experiments on a small dataset (Client 5 - 306 surveys, rules Passive to Promoter)

Test case	Running time (min)	Coverage (%)
Base	11	80.6
Type 0	0.9	74.2
Type 1	0.4	74.2
Binned—base	7.1	77.4
Binned type 0	0.4	58.1
Binned type 1	0.3	58.1

For the same type of experiment on the other small dataset (Client 5 - 306 surveys)—the results were similar (Table 7.3). Even the greater speedup was observed by switching to "Type 0" (from 11 to about 1 min) and to rule "Type 1"—about half a minute. However, the coverage slightly worsened—from 76.5% in base case to 70.6% for rule "Type 0" and 67.6% for rule "Type 1". The effect of binning the dataset was similar as in the previous small dataset (the times were better in case of binning).

7.5.2.2   Medium Datasets

For medium, in terms of size, datasets—Client 16 and Client 15 were chosen from the category Service 2016. For a medium dataset and rules "Detractor to Promoter", one could observe a speedup from the base case (Table 7.4)- from around 2 h down to about 30 min for rule "Type 0" and down to about 4 min for rule "Type 1". Running times were noticeably shorter—these kind of datasets (and larger) become problematic for the scalability of the system, as the mining on them runs for hours. Binning also shortened running times, and coverage was actually better (for the base case and rule "Type 0").

For the same case of dataset, mining rules "Passive to Promoter" was considerably longer—about 28 h in the base case (18 h for binned base case). For the new rule format, the times were about 2 h and half an hour for non-binned and 19 min and 11 min for binned datasets (see Table 7.5). The coverage slightly worsened (from targeting 153 in base case down to 143 and 138 passives). For binned datasets the

Table 7.4 A summary of experiments on a medium dataset (Client 16 - 1192 surveys, rules Detractor to Promoter)

Test case	Running time (min)	Coverage (%)
Base	169.8	86.9
Type 0	32	85.2
Type 1	4.2	82
Binned—base	131.1	88.5
Binned type 0	14.5	88.5
Binned type 1	4.9	73.8

Table 7.5 A summary of experiments on a medium dataset (Client 16 - 1192 surveys, rules Passive to Promoter)

Test case	Running time	Coverage (%)
Base	28 h	85
Type 0	2 h	79.4
Type 1	24.3 min	76.7
Binned—base	18 h	87.2
Binned type 0	19.2 min	61.1
Binned type 1	11.4 min	61.1

coverage was slightly better for base case—157 passives, but worse for rule "Type 0" and "Type 1"—110 passives (out of 180 in total).

For the second medium dataset–Client 15 with 1240 surveys in a dataset—the characteristics of the experiments were analogous. The speedup from switching to rule "Type 0" was from 10 min down to 2 min and to rule "Type 1"—further down to half a minute, for binned case—from about 8 min down to about a minute for changing rule format. The coverage was better in all cases for binned datasets. Extracting "Passive to Promoter" rules took about 11 h, for rule "Type 0"—30 min, and rule "Type 1"—10 min. For binned datasets, these times were, correspondingly: 6 h, 4 and 3 min. This time the coverage for the binned datasets was worse than for non-binned ones.

7.5.2.3 Distributed Implementation

For some datasets, the experiments on extracting action rules with Spark were conducted. The results are summarized in Table 7.6 (running times) and 7.7 (coverage). The times were measured on rule "Type 0"—that is allowing flexible attributes to only go up.

The implementation in Spark involves distributed algorithm for action rule mining. It is based on the vertical division of datasets (based on attributes)—the mining runs in a parallel way for each partition and at the end the results are combined. One could see that action rule mining run in considerably shorter times. Especially important

Table 7.6 Comparison of running times in Spark and Java for datasets from category parts 2015

Dataset	Spark (min)	Java
17–547 rows	0.9	8.6 min
16–2078 rows	1.4	3.7 h
20–2590 rows	3.0	17 h
30–3335 rows	3.0	10 h

Table 7.7 Comparison of coverage in Spark and Java for datasets from category parts 2015

Dataset	Spark (%)	Java (%)
17–547 rows	77.3	77.3
16–2078 rows	73.8	76.9
20–2590 rows	79.5	81.8
30–3335 rows	79.8	79.8

is reducing times from hours or even days (the largest dataset presented) to minutes. Also, the coverage of the algorithm in Spark is not much worse.

7.5.3   New Rule Format in RS

After testing action rule mining for running times and for the coverage of customers, the new format of rules was tested in the recommender system. Especially important was checking rules "Type 1" since they change the triggering mechanism in the system. As described in the first chapters, in the previous implementation, the triggering of action rules by meta actions (based on links between these) is based on the values ("before" and "after" values) of benchmarks. Also, binned versions need modification to the RS engine. The original and modified versions of the algorithm are depicted in Listings 1 and 2.

As one can see the algorithm for collecting all the meta actions associated with atomics of an action rule had to be modified for rule "Type 1". It is because in this type the right-hand side of an atomic containing benchmark attribute has been replaced by the value "higher". On the other hand, in the original datasets there are still original values of benchmarks, so from them it could not be discerned which rows correspond to the values "higher". In a new version, all meta-actions extracted from comments related to all the rows containing values of the benchmark higher than the left-hand side value of the atomic have to be collected. Also, there was a need to discern between binned and non-binned versions of the datasets. In the latter case all the original values had to be replaced by numbers denoting categories: 10-low, 20-medium, 30-high, 40-very high.

After adapting the triggeration mechanism of the system to work on the new type of rules and datasets, the tests were conducted on mining the meta-actions and creating meta-nodes (recommendations) based on different rule types. The following

---

Algorithm 1 Original algorithm of action rule triggeration.

1: //For each atomic action in a rule
2: **for** each atomic **do**
3:     //Parse atomic into attribute name, and "value change" part
4:     //Atomic: BenchmarkAll,(1-9)
5:     //Get attribute name
6:     $attr \leftarrow atomic.split(",")[0]$
7:     $score \leftarrow atomic.split(",")[1]$
8:     //Get "detractor score"
9:     $detractorScore \leftarrow score.split("-")[0]$
10:     //Get "promoter score"
11:     $promoterScore \leftarrow score.split("-")[1]$
12:     //Collect meta actions from rows containing the detractor's value
13:     $metaActions \leftarrow MapMetaActionstoAtomicviaforeignKeyinXLS$
        $(foreignKey, detractorCommentFilePath, attr, detractorScore,$
        $workIDMetaActions, metaActions)$
14:     //Collect meta actions from rows containing the promoter's value
15:     $metaActions \leftarrow metaActions$
        $+ MapMetaActionstoAtomicviaforeignKeyinXLS$
        $(foreignKey, promoterCommentFilePath, attr, promoterScore,$
        $workIDMetaActions, metaActions)$
16:     //Put all collected meta actions into structure
17:     $atomicMetaActions \leftarrow atomicMetaActions + (atomic, metaActions)$
18: **end for**

---

metrics have been used to keep track of the processes of the meta action mining and meta nodes creation:

A: Action rules extracted—the total number of action rules extracted with the Miner (including redundant ones),

B: Action rules read into RS—action rules that were actually used for the knowledge engine of the system (removing redundant ones and doing the reformation),

C: Ratio B/A—informs about the scale of redundancy (actions extracted versus actually used in the system),

D: Atomic actions—the total number of distinct atomic actions collected from all the action rules,

E: Atomic actions triggered—atomic actions that have associated meta actions,

F: Triggered action rules—action rules, whose all the atomics have been triggered,

G: Ratio F/B—part of all the rules read into the system that were triggered by extracted meta actions,

H: Meta actions extracted—total number of distinct meta actions extracted from the customers' comments,

I: Effective meta-nodes—meta nodes created by combining different meta actions to get the best effect on NPS,

J: Max NPS Impact—the maximal NPS Impact one can get considering all the meta-nodes (the NPS Impact of the best meta-node).

Algorithm 2 Modified algorithm of action rule triggeration to work on rule Type 1 and binned datasets.

1: //For each atomic action in a rule
2: **for** each atomic **do**
3:    //Parse atomic into attribute name, and "value change" part
4:    //Atomic: BenchmarkAll...,(1-9)
5:    //Get attribute name
6:    $attr \leftarrow atomic.split('','')[0]$
7:    $score \leftarrow atomic.split('','')[1]$
8:    //Get "detractor score"
9:    $detractorScore \leftarrow score.split(''-''[0])$
10:   //Get "promoter score"
11:   $promoterScore \leftarrow score.split(''-''[1])$
12:   //Case for rule Type 0 or stable attribute works the same as before
13:   **if** $ruleType == 0$ **or** $detractorScore == promoterScore$ **then**
14:     //Collect meta actions from rows containing the detractor's value
15:     $metaActions \leftarrow MapMetaActionstoAtomicviaforeignKeyinXLS$
16:     $(foreignKey, detractorCommentFilePath, attr, detractorScore,$
17:     $workIDmetaActions, metaActions)$
18:     //Collect meta actions from rows containing the promoter's value
19:     $metaActions \leftarrow metaActions+$
20:     $MapMetaActionstoAtomicviaforeignKeyinXLS(foreignKey,$
21:     $promoterCommentFilePath, attr, promoterScore, workIDmetaActions),$
22:     $metaActions)$
23:     //Put all collected meta actions into structure
24:     $atomicMetaActions \leftarrow atomicMetaActions + (atomic, metaActions)$
25:   **else**
26:     //Case for rule Type 1
27:     //Collect meta actions from rows containing the detractor's value
28:     $metaActions \leftarrow MapMetaActionstoAtomicviaforeignKeyinXLS$
29:     $(foreignKey, detractorCommentFilePath, attr, detractorScore,$
30:     $workIDmetaActions, metaActions)$
31:     //Since the right side of flexible attribute value is "higher" we need to check all the values higher than detractor's score
32:     //Case for non-binned datasets—the detractor's score must be lower than 10
33:     **if** $detractorScore < 10$ **then**
34:       **for** $promoterScore = detractorScore + 1; promoterScore = 10$ **do**
35:         $metaActions \leftarrow metaActions+$
36:         $MapMetaActionstoAtomicviaforeignKeyinXLS(foreignKey,$
37:         $promoterCommentFilePath, attr, promoterScore,$
38:         $workIDmetaActions, metaActions)$
39:         $promoterScore \leftarrow promoterScore + 1$
40:       **end for**

Tables 7.8, 7.9 and 7.10 show results for the metrics defined above for three types of test cases that were found most promising given running time and coverage obtained in the previous experiments: rule "Type 0", rule "Type 1" and "Bin-Type 1" (that is, "Type 1" for binned datasets).

For a small dataset (Table 7.8), not many rules were extracted in general. About twice as many rules were extracted for "Type 0" than for "Type 1" and twice as

Algorithm 3 Modified algorithm—Part 2

41:     //Case for binned datasets—the values of benchmark attributes were changed to categor-
        ical numbers: 10-low, 20-medium, 30-high, 40-very high
42:     **else**
43:       **for** $promoterScore = detractorScore + 10$; $promoterScore = 40$ **do**
44:         $metaActions \leftarrow metaActions+$
45:         $MapMetaActionstoAtomicviaforeignKeyinXLS(foreignKey,$
46:         $promoterCommentFilePath, attr, promoterScore, workIDmetaActions,$
47:         $metaActions)$
48:         $promoterScore \leftarrow promoterScore + 10$
49:       **end for**
50:     **end if**
51:     //Put all collected meta actions into structure
52:     $atomicMetaActions \leftarrow atomicMetaActions + (atomic, metaActions)$
53:   **end if**
54: **end for**

Table 7.8 Comparison of meta action mining, triggeration and meta node creation processes for a small dataset with different rule types

Metric	Type 0	Type 1	Bin-Type 1
A: Action rules extracted	329	141	179
B: Action rules read into RS	102	46	46
C: Ratio B/A	31%	32.6%	25.7%
D: Atomic actions	12	8	8
E: Atomic actions triggered	12	8	8
F: Triggered action rules	102	46	46
G: Ratio: F/B	100%	100%	100%
H: Meta actions extracted	7	7	7
I: Effective meta nodes	4	1	0
J: Max NPS impact	3.21%	2.89%	0%

many were read into the system. All of the atomic actions and the action rules were triggered. There were 7 meta actions extracted in total. With "Type 0" rules, the algorithm generated 4 meta nodes, while for "Type 1" only one meta node, and for its binned version no meta node was created. Also, the maximal NPS Impact was slightly worse for rule "Type 1".

For a medium-size dataset, about five times more rules were extracted for "Type 0" than for "Type 1". For the binned version of "Type 1" there were three times more rules extracted than for the non-binned. The ratios of redundant rules were more or less the same (about one fourth of them were redundant). Similar, as for the small

Table 7.9 Comparison of meta action mining, triggeration and meta node creation processes for a medium dataset with different rule types

Metric	Type 0	Type 1	Bin-Type 1
A: Action rules extracted	24,515	5,153	14,233
B: Action rules read into RS	5,686	1,230	3,066
C: Ratio B/A	23.2%	23.9%	21.5%
D: Atomic actions	137	54	44
E: Atomic actions triggered	135	54	44
F: Triggered action rules	5,555	1,230	3,066
G: Ratio: F/B	97.7%	100%	100%
H: Meta actions extracted	9	9	9
I: Effective meta nodes	18	5	6
J: Max NPS impact	8.58%	7.37%	1.9%

Table 7.10 Comparison of meta action mining, triggeration and meta node creation processes for a large dataset with different rule types

Metric	Type 0	Type 1	Bin-Type 1
A: Action rules extracted	59,508	11,799	31,228
B: Action rules read into RS	13,423	2,624	6,121
C: Ratio B/A	22.56%	22.24%	19.6%
D: Atomic actions	286	89	55
E: Atomic actions triggered	160	69	40
F: Triggered action rules	4,879	892	1,585
G: Ratio: F/B	36.4%	34%	25.9%
H: Meta actions extracted	11	11	11
I: Effective meta nodes	49	6	4
J: Max NPS impact	3.7%	3.45%	1.65%

dataset, all the action rules read were triggered (only in the case of "Type 0" it was 97.7%). There were 9 meta actions extracted, from which the algorithm created 18 meta nodes based on rule "Type 0", 5 meta nodes for "Type 1" and 6 nodes for

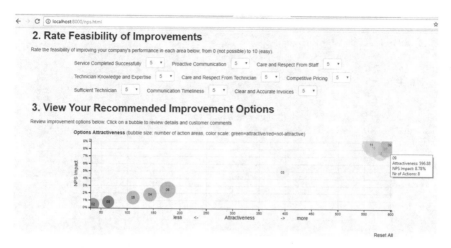

Fig. 7.1 Recommendations generated for client 32, category service, 2016 based on extracted rules of "Type 0"

"Bin-Type 1". Again, there was slightly worse maximal impact one could get for "Type 1". For the binned version, the loss on impact was significant.

The large dataset was also checked for the process of generating recommendations in the system. The statistics for action rule mining have similar characteristics between different rule types as in the previous cases. However this time, only about a third of all the action rules read into the system were actually triggered by meta actions (11 extracted in total). There were 49 meta nodes created for "Type 0", 6 for "Type 1" and 4 for "Bin-Type 1". The maximal NPS impact was comparable for "Type 0" and "Type 1" but again, much worse for the binned version.

Generated recommendations were checked in a visualization system for a chosen Client (32) from 2016. Since the binned version has not created sufficient number of nodes for the cases tested in the previous experiments, only rule "Type 0" and "Type 1" non-binned versions were checked.

From Figs. 7.1 and 7.2 one can see that using rule "Type 1" results in very limited recommendations. After testing also other cases of companies, it seems that the case is that using rule "Type 1" in the system produces meta nodes that usually contain all the extracted meta actions or eventually one or two less. It basically excludes all the nodes containing the smaller number of meta actions. This results directly from the implemented triggering mechanism. The rules of "Type 1" tend to aggregate (or target) the greater number of matched customers and differentiate between these less. This is because the right hand side of the flexible attributes was changed to "higher" and aggregates much more customers now. As a result, each action rule is associated with the greater number of customers and therefore the greater number of meta-actions extracted from the comments left by them. As a matter of fact, after doing some tests, it turned out that almost all of the action rules are associated with a complete set of extracted meta actions (extracted in general). So, the algorithm

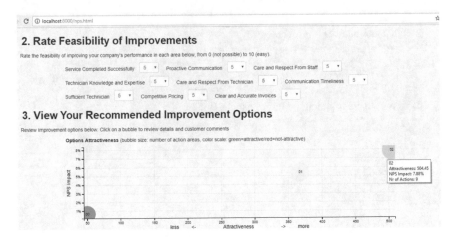

Fig. 7.2 Recommendations generated for client 32, category service, 2016 based on extracted rules of "Type 1"

for creating meta nodes, which starts with the most frequent meta actions, basically starts with creating a node containing all the meta actions (since there are no rules associated with the smaller number of meta actions).

The result for the recommender system and its user is that the system loses some of its functionality. As one can see from Fig. 7.1, one of the purposes of the recommendations is to give its users different options, depending on the needs and feasibility of particular change areas. For example, a user might want to choose a recommendation (depicted as "bubble") which is less attractive, but contains a smaller number of meta actions. In case of limited recommendations, when the system basically suggests to change all the areas which were mentioned by the customers, much of the system's functionality is lost. For example, as seen in Fig. 7.2, a user can only choose an option containing all the actions (9 in total) or eventually ones containing one or two less actions (which are shown as non-attractive)—no other options are displayed and provided. That might be especially problematic for smaller datasets, where 0 or 1 recommendation node is created.

Because of this loss of functionality of the system and lack of the solution currently, it was decided that rules of "Type 1", despite its very quick extraction time and good coverage and despite effort of adapting the system to use them in the triggering mechanism, should not be used in the system. They can be used when the time of the system's update becomes very critical or for very large datasets, when running time is unacceptable. Also, when the number of generated recommendations is varied and sufficient for a user to have a choice.

7.6 Plans for Remaining Challenges

Although the problem with long running times of action rule miner was solved and tested, there are still challenges and problems remaining to be solved, including:

- Finding a solution to make the system work with rule "Type 1" to produce a greater variety of recommendations. For example, relaxed strategy for triggeration might be applied.
- Conducting more tests—large and very large datasets chosen for tests could not be tested for all the defined cases (especially base cases and "Type 0")—as the running time cannot be accurately measured taking into account long times of running (days and weeks) and associated machine faults.
- Conducting more tests on Spark implementation and analysis of distributed algorithm—it might be that an algorithm based on "assembling" action rules from partition might not result in minimal rules. However, very short running times for extracting rules and good coverage seems to be a promising way of improving action rule mining and might become the ultimate solution for the system's scalability.
- Using other external software for extracting action rules—for example, LispMiner, which implements an efficient algorithm for mining (based on GUHA procedure) [1].

Reference

1. M. Simunek. *Academic KDD Project LISp-Miner*, pages 263–272. Springer Berlin Heidelberg, Berlin, Heidelberg, 2003.

Chapter 8
Recommender System Based on Unstructured Data

8.1 Introduction

As the time constraints for completing customer satisfaction surveys are becoming tighter, there emerges a need for developing a new format of surveying customers. The idea is to limit the number of score benchmark questions, and let customers express their opinions in a free format. As a result the collected data will mainly contain open-ended text comments.

This presents a challenge in the research, as the Customer Loyalty Improvement Recommender System (as described in Chap. 4) was built to work on both: numerical benchmark values (structured data) and text features (unstructured data). Since there will be no more numerical benchmarks in the datasets, a strategy was developed and implemented as a modified Recommender System (CLIRS2) to work on unstructured data only to make it useful with the new survey format.

8.2 Problem Statement

To remind, in the previous approach, action rules were mined from the decision tables containing benchmark scores (with Promoter Status as the decision attribute), while meta actions, triggering action rules, were mined from text comments associated with benchmarks.

8.3 Assumptions

In the previous approach, mined action rules discovered patterns as in the following example:

© Springer Nature Switzerland AG 2020
K. Tarnowska et al., *Recommender System for Improving Customer Loyalty*,
Studies in Big Data 55, https://doi.org/10.1007/978-3-030-13438-9_8

Listing 5 A sample pattern mined in the previous approach
((B1: 2 −> 5) AND (B4: 7 −> 8) AND (B9: 9 −> 10))
=> (Detractor −> Promoter)

where B1, B4 and B9 denote codes of benchmark questions. This rule is interpreted as: if the values of benchmark questions are changed from value given in the left-side to the value as given in the right side, it is expected to change customer's status from being Detractor to become Promoter. Each such rule is characterized by two metrics: support and confidence. Support of the rule is understood as the number of customers affected by this rule, that is, the greater the support the potentially more customers can be changed from Detractor to Promoter. Confidence says how probable it is to change the promoter status. Therefore, a multiplication of support and confidence (support x confidence) was developed as a new metric denoting NPS impact (expected number of customers affected) associated with an action rule.

Meta actions can be understood as extracted keywords developed into actionable statements, such as "Ensure Service Done Correctly". These were, on the other hand, mined from text data. They were also used as a way to "unify" opinions that use different wording but are semantically related to the same area (saying about the same things in different words). Meta actions were extracted from the columns with text comments, independently from mining action rules. Parsing and sentiment analysis techniques were used for meta actions extraction from raw text.

Having action rules and meta actions, triggering mechanism was implemented: meta actions, understood as actual solutions, trigger action rules, from which the projected impact on the NPS can be computed. Only these action rules were triggered (that is, included in the calculation of the NPS impact) that are associated with the comments (come from the same row in the table), from which a meta action was extracted. In this approach meta actions can trigger many action rules and the overall impact of all meta actions (overall number of customers affected) is presented as the final expected NPS Impact.

The developed recommender system CLIRS presents its output, that is, recommendable items, as collections (sets) of different meta actions (see Fig. 8.1). These sets are called meta-nodes. For each such meta node (combination of different number of meta actions) NPS Impact is calculated and shown to end user as a recommendable item. The final attractiveness of recommendation is also determined by feasibility assigned to each meta action individually. The nomenclature in relation to the developed program is shown in Fig. 8.1.

Since the new datasets of customer survey are expected to lack information on scores assigned to each benchmark, the action rules cannot be mined as described in the previous approach (see the section above). The new datasets are expected to contain mainly raw text (open-ended comments), and possibly one, two or three (obligatory) questions. Therefore, the previously built system has to be adjusted to this new open-ended survey format.

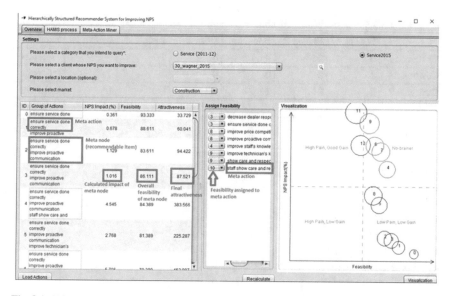

Fig. 8.1 Window interface for Recommender system presenting terms used in the initial approach

8.4   Strategy and Overall Approach

8.4.1   Data Transformation

Firstly, it is proposed to represent the data in a format suitable for the system. The
transformation of the text data needs to be done to make it suitable for data mining
(adding a structure to the unstructured data). Such new resulting data structure would
be a table, with the columns (features) as predefined keywords (called aspect classes
in the previous approach). The following "buckets" were defined (S—solely for
Service category and P—solely for Parts category):

- Service Quality General (S)
- Service Completeness
- Service Timeliness (S)
- Communication Quality
- Communication Timeliness
- Staff Attitude
- Staff Knowledgeability
- Staff Resource
- Technician Attitude (S)
- Technician Knowledgeability (S)
- Technician Resource (S)
- Invoice Accuracy

Table 8.1 Example of decision table built from expected opinion mining

	Bucket 1	Bucket 2	Bucket 3	Bucket 4	Bucket …	Bucket N	Status
Customer 1	0	2	0	0	1	−1	Promoter
Customer 2	2	2	0	0	1	0	Passive
Customer 3	0	0	1	−1	0	1	Detractor
Customer N	…	…	…	…	…	…	Promoter

- Invoice Expectation
- Invoice Timeliness
- Price Competitiveness
- Parts Availability (P)
- Parts Quality (P)
- Order Accuracy (P)
- Ease of Order (P)
- Order Timeliness (P)

The new textual data is expected to contain as much information as possible on each of these aspects. To ensure this, it is proposed to incorporate "prompt" mechanism in the survey system—prompting the surveyed customers to express opinions regarding certain topics.

As a result of data transformation, from now on, each survey (customer) in a decision table will be represented as an N-vector of values from the range $< -2; 2 >$, where N denotes the number of aspect categories, as defined above (Table 8.1).

8.4.2 Action Rule Mining

From the prepared opinion decision table action rules will be mined, with action rules of the format as follows:

Listing 6 Sample action rule for a new approach.
(Service Quality: $-2 -> -1$) AND (Technician Knowledge: $-1 -> 1$) AND (Price Competitiveness: $-1 -> 2$) => (Detractor $->$ Promoter)

The attributes of the atomic action in the rule are now aspect classes, for which sentiment was mined. The rule shows sentiment polarity's change of a bucket that should be done to change a customer from being Detractor to being Promoter. The same, these buckets can be now considered actionable knowledge (as meta actions in the previous approach) and can be used for final recommendations and NPS Impact calculation (based on action rules' support and confidence). Different combinations of aspects should be checked so that the greatest number of customers with possibly greatest confidence is affected.

8.4.3 Ideas for the Improvement of Opinion Mining

For the new format of the system, the opinion mining algorithm is critical for getting as complete opinion table as possible and therefore suitable for action rule mining. The following ideas were initially proposed to improve the current algorithm (which in the previous version was used for mining the meta actions, which then were used to trigger action rules):

- Introduction of new dictionaries for sentiment detection: SentiWordNet and AFINN (described in greater detail in Chap. 3, section on Text Mining). By using additional dictionaries it is expected to increase coverage of the algorithm, as potentially more words can be detected as opinions.
- Using nouns and verbs as opinion words—traditionally in most opinion mining algorithms, adjectives and adverbs are used as words for detecting opinions, however there might be cases that verbs and nouns also indicate the presence of sentiment. This should be possible by using new dictionaries—besides the current adjective lists—SentiWordNet and AFINN, which contain verbs and nouns with assigned polarity scores to them. However, this approach might create a new problem—if a word is first detected as opinion word, it might not be later considered as feature seed word (features and their aspects are usually detected by nouns or verbs).
- Increasing polarity scale from $\{-1, 0, 1\}$ to $\{-2, -1, 0, 1, 2\}$—this should differentiate more between opinions and potentially result in extracting more action rules (that would, for example, indicate the necessity of polarity change from -2 to -1). This should be enabled by introducing new dictionaries, which use scale for polarity strength, instead of just assigning negative/positive labels to them. Also, strength word detection, such as "really", "very" would be added to detect the strength of polarity.

8.4.4 Sentiment Extraction

Within the new approach, it is required not only to detect polarity of sentiment (negative or positive), but also the strength of this polarity. The value of "0" denotes here neutrality and we assume that if a particular area was not mentioned in a comment, a customer is indifferent to this area. Another problem is that some of the areas (buckets) might not be relevant in some cases of surveys—then the value for the area for the survey should be NULL. The sentiment will be detected, as previously, based on dictionary (standard and manually extended) of positive and negative opinion words. Feature classes will be detected based on previously developed seed words.

8.4.5 Polarity Calculation

A new strategy for polarity detection is proposed. In comparison with the previously built solution, motivation was to possibly extend sentiment detection capabilities with the new dictionaries, as well as adding sentiment detection based on words, which are not only adjectives: nouns, adverbs, etc. Three-step opinion detection is proposed, based on three most comprehensive dictionaries of positive and negative sentiment (described in Chap. 3):

- Hu and Liu adjective list (used previously),
- SentiWordNet,
- AFINN.

The two last sentiment dictionaries assign strength of polarity to words. The results of sentiment mining from all the dictionaries are combined to increase coverage of sentiment analysis and generate less sparse opinion table.

8.4.5.1 Polarity Calculation for SentiWordNet

Each synset (a word in a context) in SentiWordNet is assigned positivity score (PosS) and negativity score (NegS) for the given context of the word [1]. Polarity is calculated as: score = PosS-NegS. It calculates the sentiment score of each word in the thesaurus as a weighted average of the scores of its synsets. The polarity calculated based on SentiWordNet data can have continuous values from the range $< -1; 1 >$. Since we are using $< -2; 2 >$ discretized scale, the following mapping is proposed:

- $< -1; -0.5) \rightarrow -2$
- $< -0.5; -0.05) \rightarrow -1$
- $< -0.05; 0.05 > \rightarrow 0$
- $(0.05; 0.5) \rightarrow 1$
- $(0.5; 1) \rightarrow 2$

8.4.5.2 Polarity Calculation for AFINN

AFINN contains 564 positive and 964 negative words [2]. Each word in AFINN list is assigned one of the following values: $-5, -4, -3, -2, -1, 0, 1, 2, 3, 4, 5$. Since polarity scale used for the future system is $-2; -1; 0; 1; 2$ the following mapping is proposed:

- $< -5; -4 > \rightarrow -2$
- $< -3; -1 > \rightarrow -1$
- $0 \rightarrow 0$
- $< 1; 3 > \rightarrow 1$
- $< 4; 5 > \rightarrow 2$

8.5 Evaluation

8.5.1 Initial Experiments

The first experiments were conducted on the data from 2016, which contains about 80,000 records. In the standard approach comments from only one column are used (*Notes for Promoter Score*). With the new proposed approach, the text from all columns denoted as "Notes"—*Interviewer Notes, Resolution Notes, General Notes* and *Notes Benchmark* for each benchmark, so that to increase the insight and informational content. Still, about 6,000 records (out of 80,000) have no comments associated with the survey.

8.5.2 Experimental Setup

The general process of Natural Language Processing and Sentiment Analysis used in the system is described in Chap. 4 in the section on *Text Mining*. For evaluating results of improvement only a small subset of 2016 data was used - from the first available client (Client 1) involving its 70 comments (35 from Promoters and 35 from Detractors). The text column, as in the previous approach, was the column for "Notes for Promoter Score". The size of the subset was chosen by taking into account how many records can be manually annotated. The text comments were manually annotated by a human with the aspects that the comment is expressing opinion about, along with their polarity. The algorithm was evaluated with two metrics:

- Accuracy—measured as the number of correctly recognized and classified opinions by the number of all opinions extracted.
- Coverage—the number of opinions extracted divided by the total number of comments (that is 70 for the chosen test dataset).

It is assumed that human annotations are 100% accurate and recognize (cover) all opinions in the text (maximal coverage). Maximal coverage for the dataset is not necessarily 100%, as not all the comments bear opinions or opinions about the predefined aspects.

8.5.2.1 Test Cases

The base case was defined as human annotations, that is the target for machine Natural Language Understanding.

The second tested case is the current algorithm for sentiment analysis, based on adjective lists from Hu and Liu.

The third case involves SentiWordNet dictionary (both with and without nouns and verbs as opinion words) and the fourth—AFINN dictionary. The last test case is combining sentiment extraction results from all the cases.

For the base case (human recognition), the coverage of sentiment was 89% (which means that in 62 comments out of 70 at least one opinionated aspect was found by a human). The accuracy of humans is assumed to be 100%, as two people familiar with the domain worked on the annotations.

The initial coverage of currently used algorithm tested on the chosen 70 comments was only 20%—opinionated aspects were extracted from 14 comments out of 70 comments in total. It means there were 48 (62–14) comments that the algorithm failed to recognize with regard to sentiment that human did. On top of it, only 10 out of 14 opinions were accurate (accuracy = 71%).

After analyzing the algorithm not recognizing the sentiment in the comments, the modifications were introduced to increase its coverage and accuracy starting with the default algorithm. They are described in the next section in greater detail with illustrative examples.

8.5.3   Improving Sentiment Analysis Algorithm

On top of proposed improvements, as stated in the previous section of this chapter, some new improvements were proposed and implemented based on observing how the algorithm works on the real examples of text.

8.5.3.1   Verbs and Nouns as Opinion Words

Using verbs and nouns as opinion words based on polarity calculated from Senti-WordNet allowed to cover more comments and patterns. For example, the following comment was covered (based on the underlined opinion word, which is a noun): "*He stated the <u>timeliness</u> of the service and the friendly personnel. He worked with Rick from the Hannibal, MO location*". It seems that often the pattern for the answers in comments is : "*He stated + noun*", which makes using nouns as opinion words a promising solution. Also, common phrasing, such as "lack of (feature word)" can help in detecting negative opinions, since "lack" is a noun with assigned negative polarity. However, in the used dictionary, many nouns and verbs have assigned polarity of "0". Indeed most verbs and nouns do not bear any opinion about them and simply state the subject or the action in the sentence. Another problem is that, with the currently implemented text mining algorithm, when a word is recognized as an opinion word, it might not be considered in a feature keyword list anymore, which might result in actual worsening of the algorithm's coverage and accuracy. One solution to the problem would be change the algorithm to detect the feature word first, and the opinion word related to it secondly.

Another problem was that SentiWordNet does not contain entries for verbs in different forms. For example, in the comment: "... *that they are not completing everything that was needed for the service*", the word "completing" is not recognized (although it is assigned polarity). The solution here was to add in the functionality of the algorithm extracting base form of the verb first, and then look for its sentiment in the dictionary.

8.5.3.2 Polarity's Strength Detection

Another important change was to introduce detecting the strength of the polarity. It was implemented two-ways: first, by introducing dictionaries that assign scale of opinion polarity to words, secondly, by detecting words that "strengthen" the polarity of words—for example "very", "really", etc.

Examples of comments (chosen from the test 70-comment dataset) and opinionated segments extracted by the algorithm that involved strong polarity are given below:

- "*Chuck stated that he wasn't there when they did it, but the mechanic who was stated that the technician did a very good job*". → did very good job: 2
- "*Mike stated the field mechanic was very knowledgable and a great guy*". → great guy: 2
- "*He stated that the technician was very friendly and handled the difficult situation with the location of the equipment repair very well and did a great job.*"
  → technician very friendly: 2
  → did great job: 2
  → handled difficult situation: -2
- "*Paul said that the service man, Chris, was very friendly, knowledgeable and "on top of it".*" → very friendly: 2
- "*He stated Charlie Ray is a really good mechanic.*" → really good mechanic: 2

8.5.3.3 Handling Misspellings

As the data with comments is prepared manually by interviewers with time constraints, it happens that there are misspellings in the text. The following examples of sentences were not recognized as opinionated, because of misspelling in the opinion word:

- "*Mike stated the field mechanic was very knowledgable and a great guy*"
- "*Kevin said they know what they are doing and they are promt.*"

This results in a smaller coverage of the algorithm than it could be potentially with the correct versions of opinion words. One way to fix this is to implement handling misspellings with special libraries. Also, some sentiment libraries (such as AFINN and Hu and Liu adjective lists) contain entries for sentiment with the most common misspelled words. As for now, misspellings seem to be quite rare and negligible cases, so no misspelling correction was yet implemented.

8.5.3.4   Dealing with Ambiguity

Some words may cause ambiguity and depend heavily on the context and domain. For example, initially the word "good" in the context of technicians (example: *"good mechanics"*) was assigned to the aspect category: "technician attitude". However, after more careful semantic analysis of the comment, the opinion holder probably meant actually rather "technician knowledge". So, the word "good" was reassigned to this category in the context of "technician".

Another example from the analyzed test dataset was: *"He stated the guys are knowledgeable and the work was good"*. The algorithm detected the word "guys" as the keyword for staff, and the generated meta action was "staff knowledgeability"= 1. The same happened to the example: *"Mike stated the field mechanic was very knowledgable and a great guy"*, which resulted in an opinionated segment: *"great guy"*= 1. Again, the initial assignment of word "guys" was to the category of staff instead of technicians. However, taking into account colloquialisms used by customers, most likely they meant technicians by saying "guys". It resulted in lower accuracy when comparing with manual human labeling of opinions. Therefore, again, this seed word was reassigned to category "technician" to improve the accuracy of the algorithm.

8.5.3.5   Expanding Feature Dictionaries

In the proposed approach, dictionary-based feature recognition is used. It means that the detection of feature words related to the discovered opinion words is based on the predefined libraries of seed words for features and more fine-grained aspects of the features. These were developed manually by looking through the large sample of comments. However, they should be dynamically updated, as the new examples are seen and especially common occurring words related to a feature or an aspect should be added to the libraries.

For example, such common word occurring in comments is "experience": *"He stated this past experience was not great"*. Opinion here is not recognized (covered), because "experience" was not entered as a seed word for any of the categories. However, as it is a commonly used phrase to express opinion about service by customers generally, the word was added to the category: "service general", which corresponds to the meta action: "Service Quality General". Another such extremely common phrase is "do a good job", but again it is not covered as opinion by the current algorithm. As a solution, it was also assigned to the aspect category: "service general" (or alternatively: "service completeness").

Another problem is that the opinion is often expressed in relation with pronouns: "they", "he", etc. For example, the sentence: *"They work efficiently"* is not covered, because "they" is not assigned to any of the feature classes. Similarly, opinions are often expressed in relation to words, such as: "Caterpillar", "dealer", "Cat", etc. To solve this problem and increase the coverage of the algorithm, all these words were added to the feature category "service". Also, the word "personnel" was added to the feature category "staff".

Besides pronouns, also names of specific technicians are used, for example: "*Paul said that the service man, <u>Chris</u>, was very friendly, knowledgeable and 'on top of it*'". Here, positive opinion words about the technician were detected, however could not be associated with any of the features (as the keyword here is "Chris"). As for now, there is no solution for this, as it would require adding all the names existing in the comments. One solution would be to use some local context for words, for example, associating "knowledgeable" with the feature "technician".

Another problem discovered while testing was that some seed words are detected because they are defined in the feature categories, but they do not result in generating any meta action. This is because they were not defined in any aspect of the feature. Example printout:

subSegments features after aggregation: (stated machine new = machine,service, having condition fixed =, they already having = having, fixed on unit = fixed,service) segment orientation:stated machine new: 1.

No meta action is generated from this example as the seed word "*machine*" for the feature "*service*" was not defined in any aspect of the service. The same case was for the segment: "*Stated that they provide fast service*"—it results in the opinionated segment: "*provide fast service: 1*", but not in any meta action, as "fast" was not defined as a seed word for the category "service timeliness". This way, not all opinionated segments become meta actions. This can be fixed manually by looking at the words that exist in the feature's seed list but are not assigned to any of the categories in the aspects.

8.5.3.6 Correcting Other Fallacies of the Algorithm

On top of the mentioned improvements, some other fallacies of the algorithm were detected and corrected.

For example, in SentiWordNet dictionary semi-supervised method was used to label words with sentiment [1]. According to the conducted tests within this research, some sentiment scores are not that accurately calculated. For example, the word "good" has a calculated weighted polarity score of 0.63 (which is mapped to 2 in the used scale), while word "great"—0.25, which after mapping corresponds to 1. As a result, the word "good" is stronger positive than the word "great". This inconsistency which occurs when using only SentiWordNet was solved by changing manually the entry in the dictionary for the word "good". The PosScore was changed from 0.75 to 0.25. The same case was with the adverb "well" (its mapped sentiment based on SentiWordNet was 2).

As mentioned previously, the dictionaries for the feature and aspect seed words can be dynamically updated. The same refers to categories themselves: they can be added or redefined. For example, one of the changes within tests was adding the following aspects for service: "Service General"—with the name for meta action: "Service Quality General" and "Service Timeliness" ("Service Completed Timely").

The last but not least, it was observed that there are opinions about parts in the service surveys, for example: *"He stated parts availability was the problem "*, *"He stated this location needs more parts available on the shelf"*, *"Parts purchase was trouble free"*. However, there are separate dictionaries defined for both categories of surveys: "Service" and "Parts". The former does not contain feature and aspect libraries for parts. Therefore, although these comments contain opinions, they are not related to any of the feature words, as the comments were given in the service surveys.

Also, in the last example, "trouble free" was not correctly classified in terms of polarity—"trouble" was detected as a negative opinion word. Therefore, another idea is to handle phrases "adjective + free" as the actual negation.

8.5.4 *Experimental Results*

After introducing changes one by one to the algorithm and dictionaries as described above, there were improvements in the accuracy and the coverage of the comments.

In general, four dictionaries of opinion words were compared between themselves, as well as with the base case of human sentiment recognition. "Adjective list" denotes the current dictionary that the algorithm is using. SentiWordNet was tested in two cases: with and without using nouns and verbs as opinion words. After comparing and analyzing results from different dictionaries separately, the final strategy for the algorithm was determined and implemented.

The results are summarized in Table 8.2. At first attempt, the coverage was calculated this way, that a comment was covered if at least one opinion was discovered

Table 8.2 Comparing results of different approaches to sentiment analysis—metrics calculated per comment

Metric	Human (%)	Adj list (%)	Senti-V/NN (%)	Senti + V/NN (%)	AFINN (%)	All (%)
Coverage-before	89	20	16	19	11	
Coverage-after	89	39	31	39	30	43
Accuracy-before	100	71	73	46	63	
Accuracy-after	100	93	95	81	100	83
Weighted metric	95	66	63	60	65	63

Table 8.3 Comparing results of different approaches to sentiment analysis—metrics calculated per opinion

Metric	Human (%)	Adj list (%)	Senti − V/NN (%)	Senti + V/NN (%)	AFINN (%)	All (%)
Coverage	99	36	28	33	24	38
Accuracy	100	92	96	88	100	95

from it. Coverage and accuracy were measured before and after introducing improvements to the algorithm and dictionaries (as described in the section above). For the final assessment of an approach, a weighted metric is used, considering the accuracy and the coverage scores equally. That is, each of these two scores is assigned the weight of 0.5—it is assumed both coverage and accuracy of the sentiment analysis algorithm are equally important for the application. As one can see from Table 8.2, the initial results for the current algorithm based on "adjective list" was highly unsatisfactory—it covers only 20% of comments versus 89% of human recognition. Also, accuracy was not satisfactory—71%. It turned out that using different dictionaries (SentiWordNet, AFINN) was not better. However, the changes in the algorithm and dictionaries for features' and aspects' keywords (as described in the section above) allowed to cover twice as many comments as before. Also, using a polarity scale helped to differentiate between strength of opinions' polarities. The accuracy improved from 71 to 93% for the approach based on "adjective list". Comparing different approaches, AFINN proved to have the worst coverage, but it was very accurate. The adjective list and SentiWordNet with nouns and verbs as opinion words have similar coverage—39%, but the latter has worse accuracy. Therefore, its weighted metric is only 60 versus 66% for adjective list. The approach based on SentiWordNet without nouns and verbs covered less (31%), but was more accurate than the version with verbs and nouns as opinion words (95 versus 81%). The final conclusion for the strategy combining all three dictionaries was that the opinion words should be firstly checked in adjective lists, secondly in AFINN, and thirdly (if not found in any of these) in SentiWordNet—the last case would be to check for verbs and nouns as opinion words. The strategy was implemented and the modified algorithm is described in the next section. The combined results yielded 43% coverage and 83% accuracy.

Additionally, experiments were also measured by means of separate opinions, not comments. That is coverage was calculated as number of detected opinions in general (not only per comment) divided by all opinions detected by a human. Accuracy was also calculated per each extracted aspect (meta-action), not per comment. The results are summarized in Table 8.3. Calculating coverage per separate opinion the results for the combined approach was about 38 with 95% accuracy. Again, the adjective list proved to be the best dictionary for opinion words (36% coverage and 92% accuracy). The second best in terms of coverage was SentiWordNet with verbs and nouns as opinion words (33%), but at the same time worse in accuracy than adjective list (88 versus 92%), Senti without verbs and nouns (96%) and AFINN (95%). After

Table 8.4 Improving sentiment analysis further for the combined approach

Metric	Human (%)	All—per comment (%)	All—per opinion (%)
Coverage	89/99	57	48
Accuracy	100	88	96

the second round of improvements (adjustments in dictionaries and the algorithm), coverage and accuracy (both per comment and per opinion) were measured again, on the final overall approach. The results are shown in Table 8.4. It was possible to increase coverage to 57% (as calculated per comment)/48% (as calculated per opinion). However, the accuracy dropped a little bit as calculated per comment (88%), but increased as calculated per opinion (96%). Overall, the algorithm was improved (the new weighted metric increased from 63 to 72%—per comment and from 67 to 72%—per opinion).

Concluding experiments, introducing modifications to the algorithm and adjusting the dictionaries helped increase coverage of comments by about twice at the first attempt and three times at the second attempt in relation to the initial coverage. It also improved the accuracy and overall quality of the sentiment analysis algorithm. However, there is still room for improvement when looking at the metrics for human recognition. New improvement ideas will be discovered in the further work of this research and more tests will be conducted to determine which opinion patterns are not detected by the algorithm.

8.5.5 Modified Algorithm for Opinion Mining

The modified algorithm for opinion mining algorithm involving NLP techniques, modified, among others, for detecting the strength of polarity, is described by the following steps:

1. Preparing the file with text comments (XLS or XLSX). The mined text might be concatenation from all the "Notes" columns or might be just the column for "Notes for Promoter Score".
2. The file containing text columns is preprocessed—file reader iterates through rows in the spreadsheet.
3. Processing the current comment, which may contain many sentences.
4. Processing the current sentence using Stanford Parser [3] (Treebank Language Pack).
5. Tagging the words in the current sentence with the Part-of-Speech labels (Stanford POS tagger [3]).
6. Creating dependency list (grammatical dependency relations based on predefined templates in Stanford package [3], *GrammaticalStructureFactory*).

7. Identifying opinion words in a sentence, using: opinion word lists (Hu and Liu/SentiWordNet/AFINN), negations list ("not", "neither", etc.), conjunctive words lists ("and", "but", "therefore", etc.), strong words list ("really", "very", "much", etc.), strong positive ("best", "great", "excellent", etc.) and strong negative words lists (worst):

   a. Check if the current word is a negation (if yes, set index for negation word).
   b. Check if the current word is a conjunction (if yes, set index for conjunction word).
   c. Check if the word is present in strong opinion words (if yes, set index for conjunction word).
   d. Check if a word is in the strong positive/strong negative list, set the polarity accordingly to 2 or −2 and change polarity accordingly for the cases of negation or conjunction.
   e. Check if the word is present in positive/negative adjective list (Hu and Liu):
      • Set polarity accordingly to the list (+1 if the word found in positive adjective list or −1 if in negative adjective list).
      • Consider cases of negation, comparative forms of adjectives and conjunction to change the original polarity.
      • If valid strength opinion word found in relation to the adjective, increase polarity strength to +2 or −2.
   f. If the word is not found in adjective lists, look for its presence in AFINN dictionary:
      • If a word found in AFINN, retrieve its polarity and use mapping to use scale $< -2; 2 >$.
      • Consider cases of negation, comparative forms of adjectives and conjunction to change the original polarity.
      • If valid strength opinion word found in relation to the adjective, increase polarity strength to +2 or −2 (if the previous polarity was +1 or −1).
   g. If a word not found in adjective lists nor in AFINN, look for it in SentiWordNet dictionary:
      • If a word found in SentiWordNet, case for adjectives and adverbs.
      • Retrieve the polarity from SentiWordNet dictionary considering the word POS tag, and use mapping function to convert the continuous numbers from the scale $< -1; 1 >$ to the scale $< -2; 2 >$.
      • Consider cases of negation, comparative forms of adjectives and adverbs and conjunction to change the original polarity.
      • If a valid strength opinion word found in relation to the adjective, increase polarity strength to +2 or −2 (if the previous polarity was +1 or −1).
      • Case for verbs and nouns as opinion words: retrieve the polarity from the SentiWordNet using the POS tag and map to the scale. For verbs, find its base form—and look in dictionary for all its possible base forms. Consider cases of negation, conjunction and strength words to change the original polarity.

8. Opinion sentence summarization—finding words related to the found opinion word, using dependency relations.
9. Finding feature keyword related to the opinion word based on the previous step's results.
10. Assigning the segment to the feature category (feature aggregation).
11. Summarizing results—grouping segments by orientation, feature classes.
12. Generating meta actions from oriented segments.

8.5.6 Comparing Recommendations with the Previous Approach

After introducing changes to the sentiment algorithm with the primary purpose of increasing coverage to decrease sparsity of *opinion table* as introduced in the section on "Data Transformation". The table is further used to mine action rules, from which recommendations of the system are built directly. The better the coverage of action rules the better potential NPS Impact as calculated for each recommendation.

8.5.6.1 Sparsity

The sparsity, calculated as the number of cells in the opinion table for which aspect-based opinion was extracted divided by the total number of cells, was about 1% initially, which denotes very high sparsity. After modifications to the sentiment analysis algorithm the sparsity shrank to about 3%. The details of the comparison of sparsity before (-b) and after (-a) modifications (initial -a) and second modifications (-a2) are shown in Tables 8.5 and 8.6—for Clients: 16 and 3. *Sp* denotes *Sparsity* calculated by dividing by the number of rows in the table, *SpT*—denotes *Sparsity Total*—that is, sparsity calculated by dividing by the total number of cells in the table. Tables show results for each possible polarity score (for "before" state there were only two polarities: 1 and -1) and for all the polarity values: "All". The percentage in the cells of the tables denotes relative occurrence of extracted opinions with the given polarity.

From the tables with the results it can be seen that sparsity of opinion tables for the chosen test cases of Client16 and Client3 was reduced significantly (about three times).

8.5.6.2 Rule Mining From Opinion Table

For the sample tested clients—Client 3 and Client 16, the initial coverage of action rule mining (coverage understood as the number of distinct customers covered), was only 10.43 and 8.2% correspondingly. When dicsarding "0" as the default value (that

Table 8.5 Sparsity of opinion table before and after modifications of the opinion mining algorithm—data for Client 16

Polarity	Sp-b (%)	Sp-a (%)	Sp-a2 (%)	SpT-b (%)	SpT-a (%)	SpT-a2 (%)
−2	0	0	4	0	0	0
−1	0.9	2.9	4	0.1	0.2	0.3
1	12.3	30.5	42.5	0.8	2.0	2.8
2	0	0	4	0	0	0
All	13.2	33.4	46.9	0.9	2.2	3.1

Table 8.6 Sparsity of opinion table before and after modifications of the opinion mining algorithm—data for Client 3

Polarity	Sp-b (%)	Sp-a (%)	Sp-a2 (%)	SpT-b (%)	SpT-a (%)	SpT-a2 (%)
−2	0	0.1	0.1	0	0.01	0.01
−1	0.9	1.6	2.5	0.06	0.11	0.17
1	11.2	30.3	44.1	0.7	2.0	2.9
2	0	0.1	4	0	0.01	0.01
All	12.0	32.0	46.8	0.8	2.1	3.1

is when no meta action was detected), and leaving only "1" and "−1" the coverage was even worse—4.92 and 7.83% correspondingly. In comparison, the coverage of action rules from the previous approach (mining from the benchmark table) was 85.2% for these datasets.

After introducing modifications to the opinion mining algorithm, the action rule mining from the resulting opinion tables were repeated to see the gain in the resulting rules' coverage of the customers. The results from action rule mining on the opinion table before and after modifications (initial—*after* and later modifications—*after2*) are shown in Table 8.7. Action rule mining was performed for two cases: with "0" as default values (when opinion was not extracted from the comment on the aspect) and without "0". Test cases with—*ben* denote action rule mining performed in the previous approach, on the benchmark tables.

As with the case of testing sparsity, the coverage after two rounds of modifications was improved to about 15%, which is, however, still far less than the coverage of rules extracted from the benchmark tables.

8.5.6.3 Generating Meta Nodes

As presented in the first sections, recommendations (called *meta nodes* in the previous approach) in the new approach are generated directly from the action rules extracted from opinion tables for each client. Previously, meta nodes were built by combining meta actions by adding one at a time. In the new approach, meta actions, that is a result

Table 8.7 Results of the action rule mining on the opinion table for Client 3 and Client 16 before and after modifications to the opinion mining algorithm

Test case	Time	Nr rules	Coverage	Coverage (%)
Client3-ben-typ1	50 min	14,769	98/115	85.2
Client3-ben-typ0	7 h	93,472	98/115	85.2
Client3-before-no0	12 s	26	9/115	7.83
Client3-after-no0	14 s	29	9/115	7.83
Client3-after2-no0	19 s	155	17/115	14.78
Client3-before-with0	1.5 min	348	12/115	10.43
Client3-after-with0	13 min	569	11/115	9.57
Client3-after2-with0	6 min	2,123	18/115	15.65
Client16-ben-typ1	4 min	5,153	50/61	85.2
Client16-ben-typ0	32 min	24,515	52/61	85.2
Client16-before-no0	4 s	113	3/61	4.92
Client16-after-no0	5 s	254	6/61	9.84
Client16-after2-no0	6 s	288	8/61	13.11
Client16-before-with0	23 s	1,181	5/61	8.2
Client16-after-with0	54 s	4,023	6/61	9.84
Client16-after2-with0	1.5 min	5,596	10/61	16.39

of text mining, are part of atomic actions in action rules. Therefore, instead of iterating through most frequent meta actions to build meta nodes, now the algorithm iterates through most frequent atomic actions extracted from action rules. The strategy for building meta nodes in a tree-like structure does not change. The new algorithm is presented in Algorithm 2.

As one can see, the difference in algorithms lies in changing the order of certain steps of the algorithm. For example, in the previous version, action rule mining preceded the text mining (meta action mining). In the new version, text mining is done first, which then serves data transformation, called "opinion table" (the structure is discussed in the previous sections). Only then, action rule mining is run on the opinion table. The previous links between action rules and meta actions which created a base for triggering mechanism is now replaced by finding direct associations between action rules and meta actions, since now meta actions are attributes in the rules' atomic actions. Based on these associations, meta nodes are built, using strategy as previously used. Starting from the most frequent meta actions and adding the next one at a time with checking the increase in NPS impact of the newly created node.

Finally, experiments have been conducted with the use of newly created opinion tables, action rule mining on the opinion tables and generating recommendations (meta nodes) directly based on action rules. The results are presented in Table 8.8.

It can be seen that the maximal NPS impact one could get from the recommendations was much higher in the previous approach—7–8% versus less than 1% for Client 16 and 4–5% versus about 0.37% for Client 3. For the version without "0" (denoted

---

Algorithm 1 Original algorithm of generating meta nodes

1: //Set current client's ID and expand the client with HAMIS procedure
2: $clientID \leftarrow panel.selectedclient$
3: $clientIDmerged \leftarrow HAMIStask().expNode$
4: //Reformat the action rules extracted for the current client and its extensions
5: $ruleset \leftarrow ActionRuleReformator(actionRulePath, clientIDmerged)$
6: //Explore the action rule set and collect all existing atomic actions.
7: $atomics \leftarrow actionRulesExploration().AtomicCollector(ruleset)$
8: **for** clientName: clientIDmerged **do**
9:    //Mine the meta-actions for each atomic action using comments in each file.
10:    $listOfResultp \leftarrow miner.MiningProcessOfSegments$
11:    $(promoterCommentFilePath,$
12:    $clientName, tagger, parser, positiveseed, negativeseed, featureclasses)$
13:    $listOfResultd \leftarrow miner.MiningProcessOfSegments$
14:    $(detractorCommentFilePath,$
15:    $clientName, tagger, parser, positiveseed, negativeseed, featureclasses)$
16:    //Trigger action rules.
17:    $atomicmetaActions \leftarrow actionRulesTriggeration().atomicActionTrigger$
18:    $(promoterCommentFilePath, detractorCommentFilePath,$
19:    $workIDmetaActions, atomics)$
20:    $rulemetaActions \leftarrow actionRulesTriggeration().actionRulesTrigger$
21:    $(atomicmetaActions, ruleset)$
22:    //Group meta-actions and find the best groups of meta-actions.
23:    $metaNode[]T \leftarrow metaActionsBooster().metaActionsOrganizer$
24:    $(clientDataFile, rulemetaActions, ruleset)$
25: **end for**

Table 8.8 Comparison of recommendations results in the previous and the new (text-only) approach for Client 16 and Client 3

Test case	Rules read	Meta action extracted	Effectve meta nodes	Max NPS impact (%)
16-ben-Typ1	1230	9	5	7.37
16-ben-Typ0	5686	9	18	8.58
16-before	302	8	31	0.5
16-before-no0	34	2	2	0.17
16-after-no0	72	5	4	0.34
16-after	906	8	46	0.84
16-after2	1272	10	89	1.34
3-ben-Typ1	3364	11	5	4.5
3-ben-Typ0	21393	11	18	4.8
3-before	111	7	28	0.37
3-before-no0	12	3	1	0.37
3-after-no0	12	4	1	0.37
3-after	186	11	32	0.74
3-after2	564	12	40	0.37

Algorithm 2 Modified algorithm of generating meta nodes

1: //Set current client's ID and expand the client with HAMIS procedure
2: $clientID \leftarrow panel.selectedclient$
3: $clientIDmerged \leftarrow HAMIStask().expNode$
4: **for** clientName: clientIDmerged **do**
5:    //Mine the meta-actions for each atomic action using comments in each file.
6:    $listOfResultp \leftarrow miner.MiningProcessOfSegments$
7:    $(promoterCommentFilePath,$
8:    $clientName, tagger, parser, positiveseed, negativeseed, featureclasses)$
9:    $listOfResultd \leftarrow miner.MiningProcessOfSegments$
10:    $(detractorCommentFilePath,$
11:    $clientName, tagger, parser, positiveseed, negativeseed, featureclasses)$
12: **end for**
13: //Build the opinion table .
14: $buildOpinionTable(actionRulePath, clientIDmerged, clientNames,$
15:    $promoterCommentFilePath, detractorCommentFilePath, workIDMetas)$
16: //Reformat the action rules extracted for the current client and its extensions from the opinion
     table
17: $ruleset \leftarrow ActionRuleReformator(actionRulePath, clientIDmerged)$
18: //Explore the action rule set and collect all existing atomic actions.
19: $atomics \leftarrow actionRulesExploration().AtomicCollector(ruleset)$
20: //Iterate through rules to find meta actions associated with each rule
21: **for** rule: ruleset **do**
22:    **for** atomic: rule **do**
23:       **if** $metaActioninatomic$ **then**
24:          $metas.add(metaAction)$
25:       **end if**
26:    **end for**
27:    $ruleMetas.put(rule, metas)$
28: **end for**
29: //Group meta-actions and find the best groups of meta-actions.
30: $metaNode[]T \leftarrow metaActionsBooster().metaActionsOrganizer$
31: $(clientDataFile, ruleMetas, ruleset)$

---

as—*no0* in the table) the difference was even greater. By introducing changes to the sentiment analysis algorithm (denoted in table as—*after* and *after2*) and thus reducing the sparsity of opinion table from which action rules are mined, the recommendations were improved (by means of number of recommendations and maximal NPS impact one could get from the best node). However, the value of maximal NPS impact was still far from the one that could be achieved with the previous approach by mining benchmark tables.

8.6   Plans for Remaining Challenges

Despite the introduced change and improvements the coverage is still unsatisfying. Particularly, the maximal NPS impact calculated from action rules' support and confidence is not that significant as one could expect. This results from the poor coverage of customers with action rules. To increase coverage, further improvements to the sentiment analysis algorithm should be introduced. This section lists remaining challenges in this area and cases of text opinions that currently are not handled by the computer machine sentiment recognition.

Generally, all the opinions that do not fall into algorithmic pattern of "opinion word + feature aspect" are not extracted, These are:

- Expressing opinions by describing the situation/incident ("storytelling") without using actual opinion words—humans can infer the sentiment from the story.
- Implicit opinions—similar as the case above, but more general, for example, humans can use objective words like numbers to express opinions.
- Complex and comparative sentences—limitations lie in the used syntactical dependencies recognition.
- Opinion and feature in one word, for example "repaired".
- Using opinion words for the purpose other than expressing opinions, for example for expressing desired state or expectations.

All the unhandled cases are described in greater details in the following subsections.

8.6.1 Complex and Comparative Sentences

The algorithm is limited to syntactical dependencies that can be recognized. It happens, that some syntactical relations are not recognized, especially in complex and comparative sentences. For example, in the sentence "...*he had an issue with the time that was charged on his invoice*", "issue" is not recognized together with the keyword "invoice". It is because of the complexity of the sentence, where the opinion is expressed in a different part of the sentence than the subject of the opinion. Here, algorithm makes a mistake and associates "issue" with "time". Customers often use expressions such as "the issue is...", "the problem is..." ("*Doug said the biggest issue he has is their sense of urgency*", "*The invoicing and prices were an issue as well*", "*Bill stated that there have been a lot of communication issues*", "*He stated parts availability was the problem*"). One solution to cover such opinions would be by adding words "issue" and "problem" as a negative opinion word in the context of the feature they are associated with.

Also, comparative sentences such as: *Matt stated that he was invoiced for more than the price he was quoted for this service*, proved to be quite problematic to detect algorithmically. Besides, "have an issue with invoice" is also an example of using a phrase to express opinion implicitly.

Another issue, that turned out to be solved, was the case of two opinion words closely together, such as in: *prompt and courteous service*. Initially, the algorithm considered the word to be opinion word only if the two previous words were not opinion words. However, as in the case described above, two opinion words connected by "and" can describe two different aspects of one feature. In the example given above, the first word relates to "service timeliness", while the second opinion word to "service quality" or "staff attitude". The modification to drop the condition helped to increase the coverage of sentiment recognition further.

8.6.2 Implicit Opinions

Implicit opinions can be expressed without explicitly using opinion words. Examples of implicit opinions in the researched domain are: *"He stated that they had to make two trips and they charged him $500 , Mike said that it is four hundred dollars for Altorfer Cat just to come there to him, before they do anything, . . ."*. Here, charging a particular amount of money means negative sentiment. Another example: *"It should have not taken and hour and half to make a hose and put it on. Nick said he has seen them do it in 30 min"*. Here, using numbers is used to compare the expected and actual time of service to express dissatisfaction of "service timeliness". Also, this comment uses specific words to describe the course of actions ("to make a hose and put it on"). Using such very specific technical terms to describe service is difficult to recognize as a feature (opinion subject). Other similar examples of such comments involve: *"Tony stated that the the mechanics replaced a cylindar on there and there was parts that they could of used but they left part of it and took part of it so they could not do much with that"*. Here, by describing a situation, a customer expresses an implicit negative opinion about "technician knowledge". Another such example: *"He stated his complaint would be, why he realizes the technician has to have a work phone and he was answering it for work , it was ringing off the hook"*, says implicitly about "technician attitude" in a negative way, however this can be only inferred by a human by analyzing the situation and context. *"Willi stated that he simply calls Altorfer Cat, they arrive,"* *"a local field guy that is usually there the same day"*—in such comments it is difficult to infer algorithmically that they are positive opinions about "service timeliness".

8.6.3 Feature and Opinion in One Word

Currently, the algorithm does not handle situations when the same word is used to describe the feature as well as the opinion. As soon as the word is picked as an opinion word, other words are searched for dependencies with this word. Therefore, example comments, such as: *"He said they found a number of issues and repaired them."* *"This then allows all Altorfer to charge more than what is expected"* are not currently recognized by the algorithm. "Repaired" and "charge" both relate to features' aspects ("service completeness" and "price competitiveness"), but also bear sentiment in these contexts (positive and negative correspondingly).

Another problem is that sometimes opinions are expressed with a single noun (without using opinion words)—as an answer to the question asked to customers to receive additional comments: "What made you Detractor?" or "What could have been improved?". As a result, the often pattern of a comment is an actual answer: "Because of + noun" or just "noun". Examples in the tested dataset are: "*Jeff stated that their <u>communicatlion</u> and that fact that they kept him in the loop of things*". The possible solution would be to assign "default" polarity based on the Promoter Status label of the customer who left the comment. For example, if no opinion word was found in the sample comment above, "communication quality" would be assigned default polarity score: positive, because it was recognized as a feature word and it was expressed by a customer labelled as "Promoter". That could potentially decrease sparsity of the opinion table significantly.

8.6.4 Opinion Words in Different Context

Opinion words can be used for the purpose different than expressing sentiment towards certain aspects. For example: "*Bob said at $117 an hour for service he <u>expects a qualified technician</u> to be working on his equipment and know the problem to repair it*". In this comment "qualified technician" and "know the problem" are used to describe the desired state or expectations. Recognizing such opinions is an algorithmic error and lowers the algorithm's accuracy (they are "false positives"). Other examples include using opinion words to express needs: "*Doug said his time frame with a River boat down is he <u>needs</u> the <u>service fast</u>*".

8.6.5 Ambiguity

Sometimes it is difficult to determine algorithmically to which aspect or feature the word belongs. For example, in "*Kurt again stated it was the not being kept in the loop, and <u>the very hefty invoice</u> he received at the end of this service*", it is difficult to determine whether the opinion relates to "invoice expectations" or "price competitiveness". Also, sometimes the wrong words are chosen as a feature, for example in "*He stated that the charges for the travel <u>time</u> were very expensive*"— "time" is recognized as a keyword for "service timeliness". One solution would be to tie certain opinion words to the specific features only, for example associating "expensive" with "price". In another comment: "*He stated they have prompt and <u>courteous service</u>*"—"courteous" actually relates to the "staff" feature rather than "service", as the keyword would suggest.

8.6.6 Misspellings

For the chosen test dataset, misspellings were calculated to occur in 7% of the comments. These are the following:

- *"Mike stated the field mechanic was very knowledgable and a great guy."*
- *"Kevin said they know what they are doing and they are promt."*
- *"Eric stated that he would like better comminication."*
- *"AJ stated that they are curtious and quick to respond when service is needed."*
- *"Jeff stated that their communicatlion and that fact that they kept him in the loop of things."*

This percentage has considered misspellings to be quite common, because of the manual data entry by the interviewers and lack of misspelling checker. Therefore, the plan is to introduce in the program some kind of misspelling checker module. The misspelling checker would have to be implemented as a separate module and using external libraries for that purpose.

8.6.7 Phrases, Idiomatic and Phrasal Verbs Expressions

Another category of cases poorly handled by the algorithm is text with phrases, common slang expressions, idiomatic expressions and phrasal verbs. Examples from the tested dataset, not handled by the algorithm in opinion detection include:

- *"He stated they were right there on the spot."*
- *"Rob stated they got the machine working."*
- *"Jeff stated that they were able to get out in a timely manner."*
- *"Chris stated did work in a timely fashion."*

8.6.8 Entity Recognition From Pronouns and Names

As already mentioned in the previous sections, pronouns are often used in relation to opinion words. This problem was somehow solved by adding certain pronouns, like "they" to dictionaries ("they" was added to the general "service" feature). However this creates new ambiguities. For example, in the comment *"Dave states that they are friendly, helpful and knowledgeable"*, "they" relates more to the "staff" or the "technician" category. The same applies to the comment: *"John shared that they do fine and there is no question; they all treat him well"*. Therefore, although the algorithm's coverage increases, it becomes not that accurate at times.

Also, entities' recognition can be enhanced by adding names, as in the example: *"Paul said that the service man, Chris, was very friendly, knowledgeable and "on top of it*". The names can be retrieved from external libraries. The names of companies (dealers) can be predefined in a dictionary as well.

To summarize this chapter, a lot of enhancement have been introduced and thoroughly tested by means of tests on the text mining and analysis, as well as in relation to the new format of the recommender system that works on text data only. Also, new areas for further improvements have been identified for the future work within the research. The new format of the recommender system is designed to work solely on text feedback data and is expected to be used in the commercial settings in the future when the text data from the survey becomes reality and replaces structured or semi-structured customer surveys.

References

1. S. Baccianella, A. Esuli, and F. Sebastiani. Sentiwordnet 3.0: An enhanced lexical resource for sentiment analysis and opinion mining. In N. C. C. Chair), K. Choukri, B. Maegaard, J. Mariani, J. Odijk, S. Piperidis, M. Rosner, and D. Tapias, editors, *Proceedings of the Seventh International Conference on Language Resources and Evaluation (LREC'10)*, Valletta, Malta, may 2010. European Language Resources Association (ELRA).
2. F. A. Nielsen. A new anew: Evaluation of a word list for sentiment analysis in microblogs. *CoRR*, arXiv:1103.2903, 2011.
3. M.-C. D. Marneffe and C. D. Manning. Stanford typed dependencies manual, 2008.

Chapter 9
Customer Attrition Problem

9.1 Introduction

Customer attrition is an important problem in many industries: banking, mobile service providers, insurance companies, financial service companies etc. It was discovered [1] that on average, most US corporations lose half of their customers every five years. Another fact is that the longer a customer stays with the organization, the more profitable the customer becomes. The cost of attracting new customers is five to ten times more than retaining existing ones. Also, about 14–17% of the accounts are closed for reasons that can be controlled like price or service. Reducing the outflow of the customers by 5% can double a typical company's profit [1].

One company from the heavy equipment repair industry within this research project expressed interest in the insight of the problem, already having their own active CRM program implemented. The goal and motivation within this topic is to determine whether there are markers in the sales trends that might suggest a customer is getting ready to defect. Observing the behavior of customers who left a company, one might be able to identify customers who might leave as well.

9.2 Problem Statement

Customer attrition problem is a new subtopic that was raised by our industrial partner company. The goal here is to retain as many customers as possible. A loyal customer might be worth much more than a new customer. Many companies drive customers away with poor customer retention initiatives (or a lack of a customer retention strategy).

One client company has implemented an active customer attrition program and already had some results in this area (we will call a company Client 22 for confidentiality purposes). Therefore, the results within this work would complement or allow to compare the results of the research done by the company independently.

© Springer Nature Switzerland AG 2020
K. Tarnowska et al., *Recommender System for Improving Customer Loyalty*,
Studies in Big Data 55, https://doi.org/10.1007/978-3-030-13438-9_9

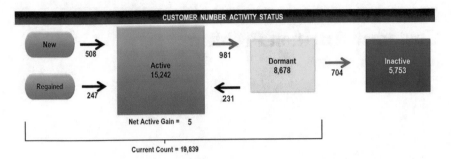

Fig. 9.1 The illustration of the data collected within the currently existing customer retention program to determine the customer activity

According to their implemented CRM program, first it is important to know if the customer base is growing or shrinking. The problem known as churn or customer attrition is defined as the annual turnover of the market base. Secondly, it is important to know which customers a company has lost recently. Targeted approaches to fight customer churn rely on identifying customers who are likely to churn, and then provide them with an incentive to stay.

The other questions that the initiative helps in answering are: How effective are we at recovering lost customers? Can we predict which customers will leave us? To answer this question, the company collects data to determine the number of customers in each activity status (see Fig. 9.1). However, the prediction capabilities of the current system are very limited. They are based on simple analysis of NLS (Net Loyalty Status) of dormant or inactive customers. Therefore, predictive capabilities are limited here as it is a simple numeric analysis based on historical data. The sample data analysis findings might include: "Even if a customer scored 9 or 10 on survey, 26% stopped doing business for a year", "37% left when they scored between 1–8". The analysis also tries to predict what impact dormant customers have on the company's business (i.e. "10–11% of Service Revenue goes Dormant each year"), taking into account the amount of business customers do. Also, the idea is to target the most lucrative customers.

The remaining questions that need to be answered using the collected data are:

- What marketing efforts should we implement to attempt recovery?
- Which marketing is having the most impact?
- Add credit line and machine ownership info to data to identify low revenue/high opportunity customers.
- How well are we cross selling?
- Breakdown by healthy customers (revenue> 1000USD)

9.3 Assumptions

The survey data from the considered company—Client 22 was available for the years 2010–2017 (total 26,454 records). Additionally, sales (transactional) data on company's customers was made available directly from that company from the years 2010–2017.

The survey (customer) data should be labelled first, in terms of customer attrition, as the current labels are only related to the *Promoter Status*. A new decision attribute, denoting whether customer left or not will be added. This can be a binary attribute (*Leaving-yes*/*Leaving-no*) or a categorical attribute *CustomerStatus*—with values: *Active, Leaving, Lost*.

As a reminder, each customer in the dataset is identifiable with a unique *Customer Number* (while there can be several contact names associated with each customer number). For each customer we can prepare temporal data for all their yearly surveys. Based on temporal records in the dataset available for that customer in the subsequent years, an algorithm can be applied for automatic labelling. The algorithm labels a customer temporal record based on the presence of that customer in the years following the survey. An additional column is added as a decision attribute.

9.4 Strategy and Overall Approach

The idea is to use survey data additionally to the sales data. This approach is novel in a way, that for the purpose of determining profile of churning customers mostly demographics or transactional data was used to mine. It is known that most customers stop doing business with a company because of bad customer service or their needs were ignored. However, customer retention efforts have also been costing organizations large amounts of resources. The customer feedback data should help in identifying and addressing the particular issues behind customer attrition. The survey data should be mined for markers of customers that are on the verge of leaving. Historical and temporal data on customer surveys should be used for that purpose. Correctly predicting that a customer is going to churn and then successfully convincing him to stay can substantially increase the revenue of a company, even if a churn prediction model produces a certain number of false positives.

The novelty of this approach lies in using transactional data to apply automatic labeling of customers in survey data as either *Active, Leaving* or *Lost*. Secondly, such automatically labelled data would be mined for actionable knowledge on how to change customer behavior from leaving to non-leaving.

9.4.1 Automatic Data Labelling

The approach lies in applying automatic labeling of customers in a survey dataset in terms of retention based on transactional data.

The first attempt to label data relied on records collected in a survey data. That is, for each survey record, the presence of the survey records for the subsequent years was checked. A customer was labelled as "Leaving-No" if there were records in the next years for that customer. A customer was labelled as "Leaving-Yes" when there were no records for that customer in the following years. However, also a so-called "Do Not Call" list must have been taken into account. Some customers wished not to be called and were entered into that list. So, if a customer was added to the list in the current or the next year, the label for the "Leaving" status could not be determined.

After comparing the previous labelling algorithm with the actual data on customer activity (in terms of invoice count) provided by the company from the sales system, it turned out that many customers were labelled as leaving although there was transactional activity recorded in the database. It was because not each transaction is followed by a telephone survey. Only a chosen subset of transactions are subject to telephone survey. So the records in the survey data are not good indicators whether a customer left or not, because the lack of records for the customer in the following years might simply result from that customer not being surveyed. Customers who were not surveyed might still have completed transactions. In conclusion, customer survey data alone cannot be used for labeling. There is a need to use additional data on the actual customer's activity—transactional data by means of invoice count in a year.

Additional data on transactions for each customer per year was made available. The revised algorithm for automatic labeling of the survey data (implemented in PL/SQL), labelled a customer as defected when: the customer appeared in the given year in the survey data, but has not reappeared at some point in the subsequent years in transactional data. As the transactional data up to 2017 was made available, survey data from years 2010–2016 could be labelled.

9.4.2 Pattern Mining

KDD (Knowledge Discovery in Databases) is defined as the "nontrivial process of identifying valid, novel, potentially useful and ultimately understandable patterns in data". The goal is to describe each customer surveyed potentially churner or potentially non-churner, so the KDD function of this problem is defined to be a classification problem. Mining patterns in the responses data would suggest likelihood to defect. The example pattern characterizing leaving customer profile is presented in Listing below.

Listing 7: A sample pattern for a leaving customer profile
IF (Benchmark1=3) AND (Benchmark2=7)
    THEN (Leaving=yes)

Action rules suggest a way how to change the values of flexible attributes to get a desired state. In the considered problem area an example suggestion to change would be the actions to decrease customer attrition. The sample actionable pattern is listed below:

Listing 8: A sample pattern mined for actionable knowledge
IF (Benchmark1 (3−>6)) AND (Benchmark2 (7−>9))
    THEN (Leaving=yes) −> (Leaving=no)

The mining results along with text mining can be incorporated into a recommender system analogically as in the previous versions of the system (but considering *Customer Attrition Status* instead of *Promoter Status*). The results should be compared with the results of Customer Attrition Program implemented by the company.

9.4.3 Sequence Mining

The available data—transactional and survey data is timestamped (date is available as an *Interview Date.*) Therefore, it is possible to detect trends in customer behavior and changes in responses over time. Sequence mining can be investigated for the prediction of customer events.

Sequence mining was originally introduced for market basket analysis where temporal relations between retail transactions are mined. The goal here is to use sequence mining for classification. If a certain sequence of responses was identified, leading to a customer defected with a high confidence, the same sequence of responses should be used for classifying customers displaying the same sequence.

9.4.4 Action Rule, Meta Action Mining and Triggering

The approach for mining the action rules, meta actions and triggering is proposed as following. Action rules showing the change in customer attrition should be mined separately for Promoters, Detractors and Passives. The next goal is to find groups of meta actions $M1$ triggering a rule $R1$ that guarantee the defection of the customers $X1$. It is assumed that if $M1$ does not happen, customers $X1$ will not defect. Analogously, find a group of meta-actions $M2$ triggering rule $R2$ that guarantee the defection of customers $X2$. As a result, for the customer group *(X1 AND X2)* the most probable set of actions causing them not to leave would be *(M1 OR M2)*.

9.5 Evaluation

The evaluation included initial data analysis, implementing and validating automatic labelling algorithm, feature selection, classification and action rules for the churn and non-churn customers. PL/SQL scripts, WEKA and LISP-Miner were used in these experiments.

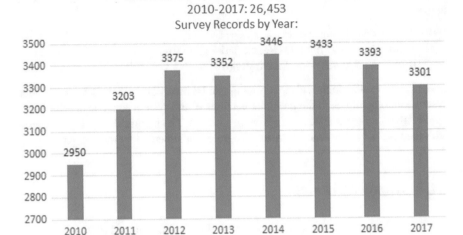

Fig. 9.2 Initial analysis of the number of surveys per year

9.5.1 Initial Data Analysis

Although about 23,000 surveys were available of the considered company (Fig. 9.2 presents the distribution among years), 8,593 distinct customers (as identifiable by *Customer Number*) were surveyed in total (Fig. 9.3). Figure 9.4 presents the distribution of the decision attribute categories resulting from applying the algorithm:

- 76% of customer surveys were labelled as "Active";
- About 10% customers left: 4% left for one year and then returned and 6% did not return at all;
- 15% of records could not be labelled.

9.5.2 Attribute Selection

Some initial experiments with data included attribute selection. Two methods within WEKA package were tested: Best First (Fig. 9.5) and Information Gain (Fig. 9.6). Both methods indicated "Overall Satisfaction", "NPS", "Likelihood to Repurchase" and general benchmarks (for the "All" category) as the most predictive in terms of attrition "Customer Status".

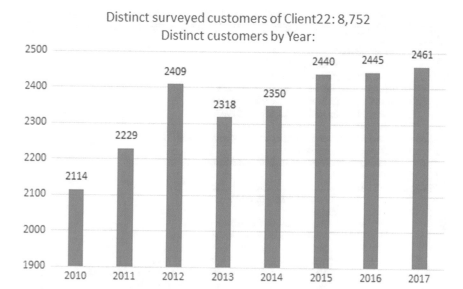

Fig. 9.3 Analysis of the number of distinct customers surveyed per year

Fig. 9.4 Analysis of the decision attribute—Customer Status—distribution of different categories

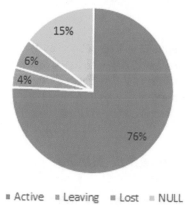

Distribution of Customer Status in Survey Data

■ Active   ■ Leaving   ■ Lost   ■ NULL

9.5.3 Classification Model

Other experiments in WEKA included building a classification method. The chosen classifier model—JRIP rules produced rules classifying customers status in terms of attrition. A sample classification rule is printed in Listing below.

Listing 9: A sample rule classifying a customer "Lost".
(Benchmark: Service − Final Invoice Matched Expectations >= 1)
AND (Likelihood to Repurchase <= 6)

```
Attribute selection output

Search Method:
        Best first.
        Start set: no attributes
        Search direction: forward
        Stale search after 5 node expansions
        Total number of subsets evaluated: 655
        Merit of best subset found:    0.005

Attribute Subset Evaluator (supervised, Class (nominal): 54 CustomerStatus):
        CFS Subset Evaluator
        Including locally predictive attributes

Selected attributes: 1,2,3,10,11,13,43,45,46,48 : 10
                     Overall Satisfaction
                     NPS
                     Likelihood to Repurchase
                     Benchmark: Dealer Promoter Score
                     Benchmark: All - Dealer Communication
                     Benchmark: Parts-How Orders Are Placed
                     Benchmark: All - Likelihood to be Repeat Customer
                     Benchmark: All - Has Issue Been Resolved
                     Benchmark: All - Contact Status of Issue
                     Benchmark: All - Contact Status of Future Needs
```

Fig. 9.5   Attribute selection using best first method

AND (Likelihood to Repurchase <= 4)
AND (Benchmark: All − Dealer Communication <= 4)
AND (Benchmark: All − Ease of Contact >= 6)
 => CustomerStatus=Lost (16.0/6.0)

9.5.4   Action Rule Mining

The actionable patterns were defined and mined in LISP-Miner (see Fig. 9.7 for the setup and Fig. 9.8 for the results). "Survey Type" was selected as a stable attribute, and all the "Benchmark" attributes were chosen as flexible attributes. The sample patterns with the highest confidence are presented in Listing below.

Listing 10: Sample rule showing reasons behind customer defection.
Status of Future Needs=1) AND (Benchmark All Ease of Contact=10) −>
(Benchmark All Contact Status of Future Needs=0) AND (Benchmark All
Ease of Contact=8) => (CustomerStatus=Active) −>
(CustomerStatus=Lost)
Survey Type(Field) : (Benchmark Service Final
Invoice Matched Expectations=9) −> Benchmark Service Final Invoice
Matched Expectations=8) => (CustomerStatus=Active) −>
(CustomerStatus=Lost)

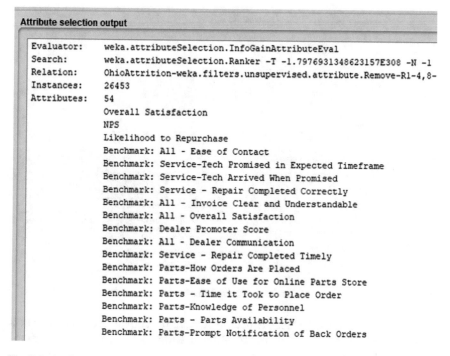

Fig. 9.6 Attribute selection using information gain method

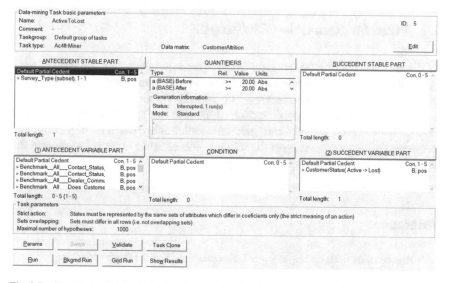

Fig. 9.7 The setup of "Data Mining" task—action rule mining with LISp-Miner to find the reasons behind the customer defection

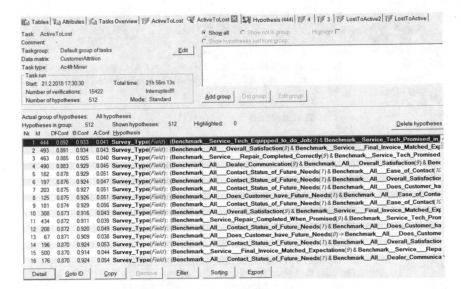

Fig. 9.8 The results of action rule mining in LISp-Miner

The found action rules associate deteriorating of the certain "Benchmark" questions, such as "Ease of Contact" or "Final Invoice Matched Expectations" with the defection of the customers.

9.6 Plans for Remaining Challenges

As the next steps within this research is to implement meta action triggering mechanism, that is, finding a minimum set of meta-actions guaranteeing that customers will not leave. The meta actions would be mined from the text in a similar way as it was for the "Promoter Status" version of the problem area. The next step would be to implement recommendations' generation mechanism based on triggering. The recommendations would suggest the minimal sets of areas to adress to prevent the largest number of customers from leaving.

Reference

1. A. Rombel. Crm shifts to data mining to keep customers. *Global Finance*, 15:97–98.

Chapter 10
Conclusions

10.1 Contribution

In this book, a data-driven user-friendly NPS-based recommender system for improving customer loyalty was presented. The first version of the system (CLIRS) was built based on both structured and unstructured data of customer feedback. The structured data was used to mine for actionable patterns and the unstructured data for the associated meta actions that act as triggers for action rules. Based on the triggering the overall impact on NPS could be calculated for different combinations of meta-actions. The second proposed version of the recommender system (CLIRS2) was built solely based on text customer feedback. The work included changing the processes (algorithm) within the first version of the system, as well as adding a step of building a numerical decision table from the text, based on detecting the numerical values of sentiment polarity towards certain aspects of the service.

The following challenges were identified within the research: (1) user-friendly interface needs to be developed to help users interact with the system and understand its output; (2) system was scaled up to 38 companies in various years—making the system fully automated required revision of data mining processes and identifying ways to improve the performance of mining; (3) text mining algorithm used for the customer comments mining was still far acceptable in terms of accuracy and coverage in sentiment mining, which was particularly troublesome when adjusting the system to work solely on text data; (4) besides customer loyalty metrics it is important to track and measure the customer churn information and try to use the survey data to identify characteristics of current customers who are likely to leave in the future. To address the above mentioned problems, the following solutions have been proposed and implemented: (a) a variety of visualization techniques used to better understand patterns in data and to visualize data mining algorithms used to generate results; interactive web-based interface to the system was developed and integrated with the system engine; (b) a variety of methods were proposed, tested and implemented in order to improve the performance of action rule mining, including: restrictions in the mined patterns, discretization of attributes and distributed data

K. Tarnowska et al., *Recommender System for Improving Customer Loyalty,*
Studies in Big Data 55, https://doi.org/10.1007/978-3-030-13438-9_10

mining (Spark); (c) a variety of methods of improving the coverage and accuracy of the opinion mining algorithm were proposed, implemented and tested, including: adding additional opinion libraries, increasing the sentiment polarity, completing feature and aspect libraries, improving entity recognition; (d) survey and transactional data for the chosen client company were combined and automatically labelled in terms of customer attrition; feature selection, data mining—classification and action rules were used to help identify customers likely to defect in the future.

10.2 Future Work

The future work includes continuation of the topic of customer attrition problem and building a new version of a recommender system in this area or a kind of "early-warning" system predicting whether the current customer is likely to defect. Furthermore, sequential pattern mining can be researched and applied in this area as this is temporal data.

Also, further work on improving the sentiment analysis should be continued: identifying new ways of improving the completeness of the decision table built out of the text. For example, assuming the sentiment based on the customer label.

Another topic involves revising the method of calculating the predicted NPS Impact so to include not only results of mining patterns showing how to convert Detractors to Promoters, but also take into account other possible changes: Detractors to Passives and Passives to Promoters. Another idea is to present the predicted overall NPS, instead of just percentage NPS change.

Another future goal is to evaluate the system in the real-world setting (with a chosen client) and compare the impact of the introduced changes as calculated by the system with the actual impact.

Printed in the United States
By Bookmasters